Lecture Notes in Computer Science 677

Habib Abdulrab Jean-Pierre Pécuchet (Eds.)

Word Equations and Related Topics

Second International Workshop, IWWERT '91
Rouen, France, October 7-9, 1991
Proceedings

Springer-Verlag
Berlin Heidelberg New York
London Paris Tokyo
Hong Kong Barcelona
Budapest

Series Editors

Gerhard Goos
Universität Karlsruhe
Postfach 69 80
Vincenz-Priessnitz-Straße 1
W-7500 Karlsruhe, FRG

Juris Hartmanis
Cornell University
Department of Computer Science
4130 Upson Hall
Ithaca, NY 14853, USA

Volume Editors

Habib Abdulrab
Jean-Pierre Pécuchet
LMAI Laboratory, INSA de Rouen
PO 08, F-76131 Mont-Saint-Aignan Cedex, France

CR Subject Classification (1991): F.4.1-3, E.1, I.2.3

ISBN 3-540-56730-5 Springer-Verlag Berlin Heidelberg New York
ISBN 0-387-56730-5 Springer-Verlag New York Berlin Heidelberg

Printed in Germany

Typesetting: Camera ready by author
45/3140-543210 - Printed on acid-free paper

Preface

This volume contains papers presented at the second International Workshop on Word Equations and Related Topics (IWWERT '91), which was held in SCUEOR, University of Rouen, from the 7th to the 9th of October 1991.

Motivated by various activities and new results in the past five years, and by the first IWWERT, which was organized in October 1990 in Tübingen, FRG, by Prof. K.U. Schulz, the contribution of the second IWWERT '91 was a very good continuation of the first one.

Several researchers working on word equations and on their applications (logic programming, automatic demonstration, system of formal calculus, combinatory of words, etc.) were present and very active at the workshop.

The workshop was chaired by Prof. G.S. Makanin whose historical contribution in this area is well known. A new research project aiming at finding a new finite description of the general solution of word equations is presented in this workshop by G.S. Makanin.

We would like to express our great thanks to LIR (Laboratoire d'Informatique de Rouen/Université de Rouen) and LMI (Laboratoire de Mathématique et Informatique/ INS de Rouen), to all the speakers and authors, and to Springer-Verlag for their optimal collaboration.

Rouen, March 1993 H. Abdulrab and J.P. Pécuchet

Contents

On general solution of equations in a free semigroup

G.S. Makanin

Steklov Mathematical Institute, Moscow

Suppose Π is a free semigroup with finite alphabet of generator

$$a_1, a_2 \ldots, a_k \qquad (1)$$

A *coefficientless* equation in Π is given by an alphabet of word variables

$$x_1, x_2 \ldots, x_n \qquad (2)$$

and a *noncancellable* equation

$$\varphi(x_1, x_2 \ldots, x_n) = \psi(x_1, x_2 \ldots, x_n) \qquad (3)$$

A list of words in the alphabet (1) $X_1, X_2 \ldots, X_n$ is called a *solution* of the equation (2), (3) whenever the words $\varphi(X_1, X_2 \ldots, X_n) = \psi(X_1, X_2 \ldots, X_n)$ coincide.

By of equation (2), (3) in Π we mean a description of all solutions of this equation by means of free word variables. In addition, the number of free word variables must not exceed n-1.

The general solution of any equation in two variables $\varphi(x_1, x_2) = \psi(x_1, x_2)$ is of the form $x_1 = (u)^\lambda$, $x_2 = (u)^\mu$, where u is a free word variable, λ and μ are either natural parameters or natural integers.

The general solution of any equation in three variables $x_1x_2x_3 = x_3x_2x_1$ is of the form $x_1 = (u_1u_2)^\lambda u_1$, $x_2 = u_2(u_1u_2)^\mu$, $x_3 = (u_1u_2)^\nu u_1$, where u_1 and u_2 are free word variables and the λ, μ, ν are natural parameters.

The general solution of any equation in three variables $\varphi(x_1, x_2, x_3) = \psi(x_1, x_2, x_3)$ is representable by a finite number of formulas which is constructed from free word variables u_1, u_2 by means of operation of multiplication and raising to a power with a variable exponent.

The general solution of the equation $x_1x_2x_3 = x_3x_4x_1$ is not representable by a finite number of formulas which is constructed from free word variables u_1, u_2, u_3 by means of operation of multiplication and raising to a power with a variable exponent.

Let $\tau = (\alpha_1, \alpha_2, \ldots, \alpha_q)$ be an arbitrary vector of integers. The integer α_q will be denoted by the symbol $\omega(\tau)$, the number q by $\upsilon(\tau)$. If $\upsilon(\tau) > 1$, the vector $(\alpha_1, \alpha_2, \ldots, \alpha_{q-1})$ will be denoted by the symbol $\mu(\tau)$.

We define here the function $[u_1, u_2, \ldots, u_p]_i^\tau$, where $p > 1$; $i = 1, 2, \ldots, p$; $u_1, u_2, \ldots, u_p$ are word variables; and τ is a vector of integers.

— If $\upsilon(\tau) = 1$, then
$[u_1, u_2, \ldots, u_p]_i^\tau \overset{\text{def}}{=} u_i$, $1 \le i \le p$.

— If $\upsilon(\tau) > 1$ and $\omega(\tau) \ge 0$, then

1) if i = 1,

$$[u_1, u_2, \ldots, u_p]_i^\tau \overset{\text{def}}{=} ([u_1, u_2, \ldots, u_p]_2^{\mu(\tau)} [u_1, u_2, \ldots, u_p]_1^{\mu(\tau)})^{\omega(\tau)+1} [u_1, u_2, \ldots, u_p]_2^{\mu(\tau)}$$

2) if $2 \le i \le p-1$

$$[u_1, u_2, \ldots, u_p]_i^\tau \overset{\text{def}}{=} [u_1, u_2, \ldots, u_p]_{i+1}^{\mu(\tau)}$$

3) if i = p,

$$[u_1, u_2, \ldots, u_p]_i^\tau \overset{\text{def}}{=} ([u_1, u_2, \ldots, u_p]_2^{\mu(\tau)} [u_1, u_2, \ldots, u_p]_1^{\mu(\tau)})^{\omega(\tau)} [u_1, u_2, \ldots, u_p]_2^{\mu(\tau)}$$

— If $\upsilon(\tau) > 1$ and $\omega(\tau) < 0$, then

1) if i = 1

$$[u_1, u_2, \ldots, u_p]_i^\tau \overset{\text{def}}{=} ([u_1, u_2, \ldots, u_p]_{p-1}^{\mu(\tau)} [u_1, u_2, \ldots, u_p]_p^{\mu(\tau)})^{-\omega(\tau)-1} [u_1, u_2, \ldots, u_p]_{p-1}^{\mu(\tau)}$$

2) if $2 \le i \le p-1$

$$[u_1, u_2, \ldots, u_p]_i^\tau \overset{\text{def}}{=} [u_1, u_2, \ldots, u_p]_{i-1}^{\mu(\tau)}$$

3) if i = p

$$[u_1, u_2, \ldots, u_p]_i^\tau \overset{\text{def}}{=} ([u_1, u_2, \ldots, u_p]_{p-1}^{\mu(\tau)} [u_1, u_2, \ldots, u_p]_p^{\mu(\tau)})^{-\omega(\tau)} [u_1, u_2, \ldots, u_p]_{p-1}^{\mu(\tau)}$$

The general solution of mirror equation in p variables

$$x_1 x_2 \ldots x_{p-1} x_p = x_p x_{p-1} \ldots x_2 x_1 \tag{4}$$

is represented by means of the following formulas

$X_i = [1, y_1, y_2, \ldots, y_{p-1}]_i^\tau$ $(i = 1, 2, \ldots, p)$ $\qquad y_1 y_2 \ldots y_{p-1} = y_{p-1} \ldots y_2 y_1$

$$X_i = [y_1, y_2, \dots, y_{p-1}, 1]_i^{\tau} \quad (i = 1, 2, \dots, p) \qquad y_1y_2 \dots y_{p-1} = y_{p-1} \dots y_2y_1 \tag{5}$$

$$X_i = [y_{p-1}, y_1, y_2, \dots, y_{p-2}, y_{p-1}]_i^{\tau} \quad (i = 1, 2, \dots, p) \qquad y_1y_2 \dots y_{p-2} = y_{p-2} \dots y_2y_1$$

Here τ is an arbitrary vector of integers, and $y_1, y_2, \dots, y_{p-1}$ are word variables connected by corresponding mirror equation with less than p variables.

Using the formulas (5) it is easy to construct a finite number of formulas describing the general solution of mirror equation (4).

By an *elementary* equation in *n* variables we mean an equation of the form

$$x_1x_2 \dots x_{n-1}x_n = x_{i_1}x_{i_2} \dots x_{i_n} \tag{6}$$

where $i_1, \dots, i_n$ is a permutation of $1, \dots, n$, and $i_1 \neq 1$, $i_n \neq n$.

For each elementary equation (6) we can construct some corresponding formulas (similar to formulas for mirror equation) which describe the general solution of equation (6).

Hypothesis

The general solution of any equation in n variables $\varphi(x_1, x_2 \dots, x_n) = \psi(x_1, x_2 \dots, x_n)$ is represented by a finite number of formulas constructed from formulas of elementary equations.

An equation (2), (3) is called *full* if both $\varphi(x_1, x_2 \dots, x_n)$ and $\psi(x_1, x_2 \dots, x_n)$ contain all letters $x_1, x_2, \dots, x_n$.

By a *directed equation* in Π, we mean an equation of the form $x_t\varphi(x_1, x_2 \dots, x_n) = x_s \psi(x_1, x_2 \dots, x_n)$ with the additional condition $\partial x_t > \partial x_s$.

This directed equation will be written as follows

$$x_t\varphi(x_1, x_2 \dots, x_n) \rightarrow x_s \psi(x_1, x_2 \dots, x_n)$$

It is easy to show that the general solution of any equation in Π is representable by means of general solutions corresponding to directed equations in Π.

Let

$$x_1\varphi(x_1, x_2 \dots, x_n) \rightarrow x_2\, \alpha(x_2, \dots, x_n)x_1\beta(x_1, x_2 \dots, x_n) \tag{7}$$

be an arbitrary directed equation in Π, (with a possible renumbering of variables).

The transformation of this equation

$$x_1 \rightarrow x_2\, \alpha_1(x_2, \dots, x_n)(\alpha_2(x_2, \dots, x_n)x_2\alpha_1(x_2 \dots, x_n))^{\lambda}\, x_1 \tag{8}$$

where $\alpha(x_2, \dots, x_n)$ is identical to $\alpha_1(x_2 \dots, x_n)\alpha_2(x_2, \dots, x_n)$, and λ is a natural parameter, is called *complete*.

The result of applying the complete transformation (8) to the equation (7) is the directed full equation

$$x_1\varphi(\tilde{x}_1, x_2 \ldots, x_n) \leftarrow \alpha_2(x_2, \ldots, x_n)x_2\alpha_1(x_2 \ldots, x_n)x_1\beta(\tilde{x}_1, x_2 \ldots, x_n) \qquad (9)$$

where $\tilde{x}_1$ is the right-hand side of transformation (8).

Starting from the equation (7) and applying the complete transformation (8), we construct a list of directed full equations $\Sigma_1, \ldots, \Sigma_{\delta\alpha+1}$ with natural parameters.

Then we will apply the complete transformation of each equation Σ_i. (Sometimes we need to replace the equation containing a component A^λ by two equations containing components 1 and AA^λ, respectively).

We will continue the construction of this "tree" of directed full equations, with natural parameters.

By a *polarizated* equation in n variables we mean any system consisting of the following four parts (with a possible renumbering of variables).

1) Polarizated alphabet

$$x_2^*, \ldots, x_n^*, x_1^+, x_2^+, \ldots, x_n^+$$

2) Directed full equations

$$x_2^*v^+\varphi(x_2^*,\ldots, x_n^*, x_1^+, x_2^+, \ldots, x_n^+) \rightarrow x_1^+\psi(x_2^*,\ldots, x_n^*, x_1^+, x_2^+, \ldots, x_n^+) \quad (10)$$

where $x_2^*v^+\varphi(\ldots) \rightarrow x_1 \ldots$ denotes

— either the equation $x_2^*\varphi(\ldots) \rightarrow x_1 \ldots$,

— or the equation $x_2^+\varphi(\ldots) \rightarrow x_1 \ldots$.

3) Polarization

A polarizated equation contains the functions $\alpha(i)$ and $\beta(i)$ with the following domains and ranges:

$\alpha(i) : \{2, \ldots, n\} \rightarrow \{1, \ldots, n\}$, $\qquad \alpha(i) = i \Rightarrow i = 2.$

$\beta(i) : \{2, \ldots, n\} \rightarrow \{1, \ldots, n\}$, $\qquad \beta(i) \neq i.$

The variables $x_2^*, \ldots, x_n^*$ are polarizated in the equation (10) by the following rule:

$$x_i^* \rightarrow x_{\alpha(i)}^* \vee x_{\alpha(i)}^*, x_{\alpha(i)}^+ \vee x_{\alpha(i)}^*, x_1^+ \vee x_{\alpha(i)}^*, x_{\alpha(i)}^+, x_1^+ \qquad (11)$$

(That is, only $x_{\alpha(i)}^*$ can be placed after x_i^* in the equation (10), or only $x_{\alpha(i)}^*$ and $x_{\alpha(i)}^+$ can be placed after x_i^*, or only $x_{\alpha(i)}^*$ and x_1^+, or only $x_{\alpha(i)}^*$ and $x_{\alpha(i)}^+$, and x_1^+).

$$x_{\beta(i)}^* \vee x_{\beta(i)}^+ \leftarrow x_i^+ \qquad (12)$$

(That is, only $x_{\beta(i)}^*$ can be placed before x_i^+ in the equation (10), or only $x_{\beta(i)}^+$ can be placed before x_i^+).

We will say that x_i^* *has* the variable v, if the polarization (11) contains $x_i^* \rightarrow v$.

We will say that x_i^+ *has* the variable v, if the polarization (12) contains $v \leftarrow x_i^+$.

4) Additional conditions

— x_i^* has x_k^+ $\Leftrightarrow$ x_k^+ has x_i^*.

— If x_i^* has x_1^+, then $\alpha(i) = 2$.

— $\forall i\, [\quad \exists t\, (x_i^*$ has $x_t^+)$,

$\vee \exists s, t\, (x_i^*$ has x_s^+ & x_s^* has $x_t^+)$,

$\vee \exists r, s, t\, (x_i^*$ has x_r^+ & x_r^* has x_s^+ & x_s^* has $x_t^+)$,

$\ldots]$

Here is an example of a polarizated equation in four variables x_2^*, x_3^*, x_4^*, x_1^+, x_2^+, x_3^+, x_4^+.

$$x_2^*\varphi(\ldots) \to x_1^+\, \psi(x_1, x_2 \ldots, x_n)$$

$$\begin{array}{ll} x_2^* \to x_4^*,\ x_4^+ & x_3^* \leftarrow x_1^+ \\ x_3^* \to x_2^*,\ x_1^+ & x_1^* \leftarrow x_2^+ \\ x_4^* \to x_3^* & x_1^* \leftarrow x_3^+ \\ & x_2^* \leftarrow x_4^+ \end{array}$$

♦

Theorem

If an equation $S(x_1, \ldots, x_n) \to T(x_1, \ldots, x_n)$ belongs to the "tree" of equations (7) and its branch contains all variables $x_1, \ldots, x_n$, then this equation is polarizated.

Remark

If the branch of the equation $S(x_1, \ldots, x_n) \to T(x_1, \ldots, x_n)$ contains the variables $x_1, \ldots, x_k$, $(k < n)$, then it is *partially polarizated*, that is its alphabet is $x_2^*, \ldots, x_k^*$, $x_1^+, x_2^+, \ldots, x_k^+, x_{k+1}, \ldots, x_n$ and its polarization is given by the following:

$$x_i^* \to \gamma_i(x_{k+1}, \ldots, x_n)x_t^*v^+ \qquad x_t^*v^+\, \delta_i(x_{k+1}, \ldots, x_n) \leftarrow x_i^+$$

♦

If the word $R(x_1, \ldots, x_n)$ contains all the letters $x_1, \ldots, x_n$, then by the *prefix* of this word, we mean a word $P(x_1, \ldots, x_n)$ of minimal length which contains all the letters $x_1, \ldots, x_n$ and such that $R(x_1, \ldots, x_n)$ is identical to $P(x_1, \ldots, x_n)\, Q(x_1, \ldots, x_n)$ for some $Q(x_1, \ldots, x_n)$.

By the *prefix-equation* of the equation $S(x_1, \ldots, x_n) \to T(x_1, \ldots, x_n)$ we mean the equation $S'(x_1, \ldots, x_n) \to T'(x_1, \ldots, x_n)$, where $S'(x_1, \ldots, x_n)$ is the prefix of $S(x_1, \ldots, x_n)$ and $T'(x_1, \ldots, x_n)$ is the prefix of $T(x_1, \ldots, x_n)$.

Theorem

The set of prefix-equations of the "tree" of equations (7) is finite and can be constructed by means of formulas of elementary equations.

♦

CONJUGACY IN FREE INVERSE MONOIDS

Christian CHOFFRUT
Laboratoire d'Informatique Théorique et de Programmation
Université Paris 7, Tour 55-56, 1er étage
2 Pl. Jussieu, Paris 75 251 Cedex 05
email: cc@litp.ibp.fr

Abstract: the notion of conjugacy in groups can be extended in two ways to monoids. We keep on calling conjugacy the first version (two elements x and y are conjugate if xz=zy holds for some z), while we call transposition the second one (two elements x and y are transposed conjugate if x=uv and y=vu holds for some u,v). Using the characterization of elements in free inverse monoids due to Munn, we show that restricted to non idempotents, the relation of conjugacy is the transitive closure of the relation of transposition. Furthermore, we show that conjugacy between two elements of a free inverse monoid can be tested in linear time.

Résumé: la notion de conjugaison dans les groupes peut être étendue de deux façons aux monoides. Nous appelons conjugaison la première version (deux éléments x et y sont conjugués si xz=zy est vrai pour un certain z), alors que nous appelons transposition la seconde version (deux éléments x et y sont transposés si x=uv et y=vu sont vrais pour certains u,v). Utilisant la caractérisation due à Munn des éléments d'un monoide inversif libre, nous montrons que restrainte aux éléments non idempotents, la première relation est la fermeture transitive de la première. De plus nous prouvons que l'on peut tester en temps linéaire si deux éléments d'un monoide libre inversif sont conjugués.

1.INTRODUCTION

Two elements x and y of a group G are conjugate if and only if there exists z in G such that $x=zyz^{-1}$. This notion is usually extended in two different ways to monoids (cf., e.g., [Os] and [Ot]). Indeed, say two elements x,y of a monoid M are *conjugate* and write Conj(x,y) if there exists $z\in M$ such that xz=zy. Say they are *transposed* and write Trans(x,y) if there exist $u,v \in M$ such that x=uv and y=vu. Clearly, conjugacy is reflexive and transitive but not necessarily symmetric. Moreover, transposition is reflexive and symmetric but not necessarily transitive. Nevertheless the following inclusions hold in all cases (where the exponents refer to the operation of composition of realations and where the star refers to the transitive closure):

$$\text{Trans} \subseteq \text{Trans}^2 \ldots \subseteq \text{Trans}^k \ldots \subseteq \text{Trans}^* \subseteq \text{Conj}$$

A few papers have dealt with comparing these two relations (essentially [LS], [Ot],[Dub] and [Zh]). Here we study the case of free inverse monoids, i.e., of the free objects in the category of regular monoids for which all idempotents commute (cf. paragraph 3.1 for a precise definition). The main purpose is to prove that in free inverse monoids, conjugacy restricted to non idempotents is the transitive closure of transposition:

Theorem *In free inverse monoids, all idempotents form a unique class of conjugacy. Moreover, two non idempotent elements* x *and* y *satisfy* xz=zy *for some element* z *if and only if there exist an integer* $n\geq 0$ *and a sequence* $x=x_0,x_1\ldots, x_n=y$ *of elements such that for* i=0,...n-1 *there exist* u,v *satisfying* $x_i=uv$ *and* $x_{i+1}=vu$.

This research was supported by the PRC Mathématique-Informatique

Also based on this result, a linear algorithm is given that tests for conjugacy in free inverse monoids. The main tool for proving this result is the characterization of the elements of free inverse monoids by ways of walks in certain trees due to Munn. This technique has been employed by other authors in connection with the solution of the word problem in some inverse monoids (cf., e.g., [Me] and [St]).

In section 2 the main definitions and notions are recalled. Section 3 is devoted to the valuable Munn's characterization of free inverse monoids and states some useful elementary properties. In section 4 the theorem is proven and complexity considerations are developed.

2. PRELIMINARIES

Given a finite nonempty alphabet Σ, we denote by Σ^* the free monoid it generates. An element of Σ^* is a *word* and the unit 1 of Σ^* is the *empty* word. Given a word $w \in \Sigma^*$, we denote by $|w|_a$ the number of occurrences of the letter a in w, and by $|w|$ its *length* : $|w| = \sum_{a \in \Sigma} |w|_a$.

A word u is a *prefix* of a word w if w=uv holds for some $v \in \Sigma^*$. Two words u,v are *comparable* if u is a prefix of v or v is a prefix of u. A word $w \in \Sigma^*$ is *primitive* if $w=u^n$ implies n=1 otherwise it is *imprimitive*. It is well-known that each non empty word is some power of a unique primitive word called its *root* (cf.,e.g., [LyS] Corollary 4.1).

Example 2.1: aba is primitive but abab is not and its root is ab□

Given a linear order on Σ we recall that it can be extended to Σ^* by defining the relation u<v if and only if:
either $|u|<|v|$
or $|u|=|v|$ and for some $x,u',v' \in \Sigma^*$, $a,b \in \Sigma$, $a<b$ we have u=xau', v=xbv'
This ordering is known as the *alphabetic* ordering.

Example 2.2: if a<b then b<ab<ba

In this work, a monoid M will be given by *generators* and *relators*, i.e., by one of its *presentations* $<\Sigma;E>$ where $E \subseteq \Sigma^* x \Sigma^*$. Alternatively, it is customary to denote any pair (u,v) of E (also called *relator* or *rule*) as u=v. Thus, e.g., <a,b; ab=ba> denotes the free commutative monoid on the two generators a,b.Thus, M is isomorphic to the quotient $\Sigma^*/_{\underset{E}{\cong}}$ where $\underset{E}{\cong}$ is the congruence over Σ^* generated by the relation E.

As said in the introduction, conjugacy in groups leads naturally to two different definitions in general monoids. Indeed, starting with the condition $x=zyz^{-1}$, either we observe that it yields the factorizations $x=(z)(yz^{-1})$ and $y=(yz^{-1})(z)$ or we multiply it by z to the right and we get: xz=zy.

In the first case we define the relation of *transposition* :

Definition 2.2.1

Two elements x,y of a monoid M are *transposed* and we write $\mathrm{Trans}(x,y)$ if and only if there exist $u,v \in M$ such that $x=uv$ and $y=vu$.

In the second case we define the relation of *conjugacy* :

Definition 2.2.2

Two elements x,y of a monoid M are *conjugate* and we write $\mathrm{Conj}(x,y)$ if and only if there exists $z\in M$ such that $xz=zy$.

Example 2.3: in free monoids the two relations of conjugacy and transposition coincide ([LS]) as well as in monoids with special presentations, i.e., with presentations where the relators are of the form $u=1$ ([Zh]). A non trivial case where the relations differ is, e.g., the free monoid generated by $\Sigma=\{a_1,a_2,\dots,a_n\}$ with the partial commutations $a_ia_j=a_ja_i$ if $|i-j|>1$, because we have $\mathrm{Trans}^{n-1}=\mathrm{Conj}$ and $\mathrm{Trans}^{n-2}\subset\mathrm{Conj}$ ([Dub], Proposition 3.12).□

3. THE FREE INVERSE MONOID

3.1 General properties

Consider $\Sigma=\{a_1,a_2,\dots,a_n\}$, a disjoint copy $\bar{\Sigma}=\{\bar{a}_1,\bar{a}_2\dots,\bar{a}_n\}$ and set $\Delta=\Sigma\cup\bar{\Sigma}$. Given any $c\in\Delta$ we define:

$$\bar{c}=\begin{cases}\bar{a}_i & \text{if } c=a_i \text{ for some } 1\le i\le n\\ a_i & \text{if } c=\bar{a}_i \text{ for some } 1\le i\le n\end{cases}$$

The application $c\rightarrow\bar{c}$ is an involution that exchanges Σ and $\bar{\Sigma}$, i.e., we have $\bar{\bar{c}}=c$ for all $c\in\Delta$. This application is extended to an arbitrary word $w\in\Delta^*$ by setting: $\bar{1}=1$ and for all $a\in\Delta$ and $u\in\Delta^*$: $\overline{ua}=\bar{a}\bar{u}$.

The *free group* generated by Σ is the monoid presented by:

$$<\Delta;F> \text{ where } F=\{\, c\bar{c}=1 \mid c\in\Delta \,\}$$

and we denote by α the canonical morphism of Δ^* onto $F(\Sigma)$.

The *free inverse monoid* generated by Σ, denoted by $\mathrm{Inv}(\Sigma)$, is presented by:

$$<\Delta;E> \text{ where } E=\{\, x\bar{x}x=x \mid x\in\Delta^* \,\} \cup \{x\bar{x}y\bar{y}=y\bar{y}x\bar{x} \mid x,y\in\Delta^*\}.$$

and we denote by β the canonical morphism of Δ^* onto $\mathrm{Inv}(\Sigma)$. Since the canonical congruence $\underset{E}{\cong}$ is finer than the canonical congruence $\underset{F}{\cong}$:

$$u \underset{E}{\cong} v \Rightarrow u \underset{F}{\cong} v$$

there exists a unique morphism π of $\mathrm{Inv}(\Sigma)$ onto $F(\Sigma)$ that makes the following diagram commute:

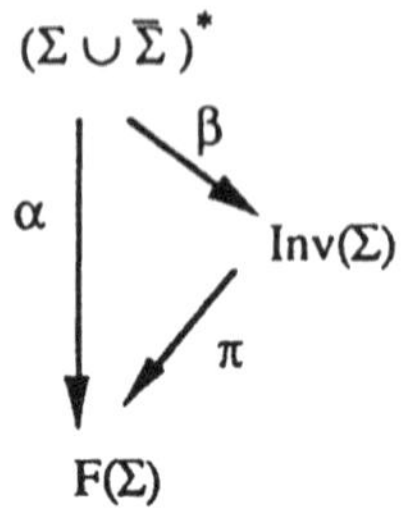

To all words $u \in \Delta^*$ we assign a *reduced* word $\rho(u) \in \Delta^*$ equivalent to u in the congruence $\underset{F}{\approxeq}$, i.e., a word that contains no factor of the form $c\bar{c}$ for some $c \in \Delta$. Inductively we have:

(i) for all $u \in \Delta \cup \{1\}$, we have $\rho(u)=u$

(ii) for all $u \in \Delta^*$, $c,d \in \Delta$, we have $\rho(ucd)=\begin{cases} \rho(uc)d & \text{if } d \neq \bar{c} \\ \rho(u) & \text{if } d=\bar{c} \end{cases}$.

We denote by $R(\Sigma)$ the set of all reduced words:

$$R(\Sigma)= \Delta^* - \Delta^*\{c\bar{c} \mid c \in \Delta\}\Delta^*$$

A word is *cyclically reduced* if it is empty or if its first letter c and its last letter d satisfy: $d \neq \bar{c}$.

Example 3.1.1: the word $ab\bar{b}a$ is not reduced. The word $\bar{a}ba\bar{b}\bar{b}a$ is reduced but not cyclically reduced. The word $\bar{a}ba\bar{b}\bar{b}ab$ is cyclically reduced. □

It is well-known that all classes of the congruence $\underset{F}{\approxeq}$ are entirely characterized by a unique reduced word (cf., e.g., [MKS], p. 183). Equivalently, the application $u \to \rho(u)$ is a bijection of $R(\Sigma)$ onto the free group $F(\Sigma)$. This leads to identifying a reduced word with its image in the free group. This identification may be extended to arbitrary subsets $X \subseteq F(\Sigma)$ by identifying X to the subset $Y=\alpha^{-1}(X) \cap R(\Sigma)$ of the reduced words of Δ^* that are representatives of the classes of X. This enables us to borrow the terminology of free monoids. In particular we say that the subset X is *prefix-closed* if $Y=\alpha^{-1}(X) \cap R(\Sigma)$ is itself prefix-closed, i.e., if all prefixes of Y belong to Y. Furthermore, the *prefix-closure* of the subset X is the image by α of the least subset of Δ^* that contains $Y=\alpha^{-1}(X) \cap R(\Sigma)$ and that is prefix-closed. It is standard to represent a subset that is prefix-closed by a tree.

What concerns the free inverse monoid, Munn proved a very useful characterization of the equivalence classes of $\underset{E}{\approxeq}$ ([Mu], Theorem 2.8. see also [S1], [S2]). Intuitively, a word $u \in \Delta^*$ can be viewed as specifying a walk in the Cayley graph $\Gamma(\Sigma)$ of the free group $F(\Sigma)$, that starts in $\alpha(1)$ and ends in $\alpha(u)$. Then the class of u

in the congruence $\underset{E}{\approxeq}$ is uniquely determined by the set T(u) of all the vertices that have been visited, no matter in which order, along with the last visited vertex $\alpha(u)$.

Formally we define T(u) inductively as follows:

(3.1.1) *if* u=1 *then* $T(u)=\{\alpha(1)\}$

(3.1.2) *for all* $u\in\Delta^*$ *and* $c\in\Delta$, $T(uc)=T(u)\cup\{\alpha(uc)\}$

If we set $I(u)=(T(u),\alpha(u))$, we get:

Theorem 3.1.1 (Munn) u $\underset{E}{\approxeq}$ v *holds if and only if* $I(u)=I(v)$

Because the Cayley graph of the free group is an (infinite) tree (cf. Fig. below), T(u) itself is a (finite) tree. Thus, the pair $I(u)=(T(u),\alpha(u))$ can be viewed as a birooted tree (cf. [Mu], p. 393).

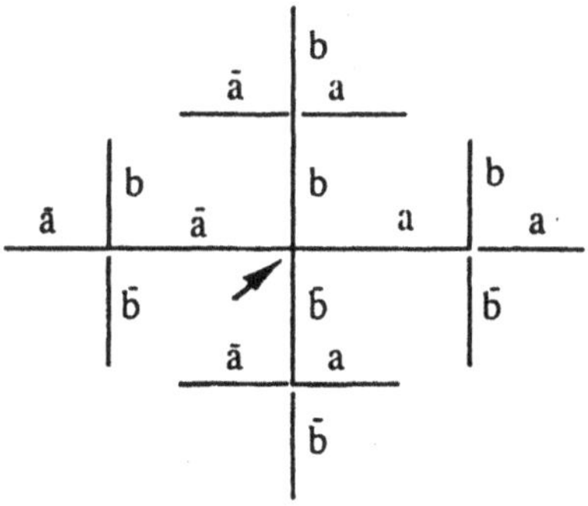

Figure 1: the Cayley graph of the free group $F(\Sigma)$

Example 3.1.2: the class of equivalence of the word $\bar{a}b\bar{b}ab\bar{b}a\bar{b}ba\bar{a}ba\bar{a}b\bar{b}$ is represented by the graph below. The origin of the walk is represented by an arrow and the last visited vertex by a circle. By convention the letters a and ā are represented by horizontal segments and the letters b and b̄ by vertical segments (the orientations of the edges are omitted).

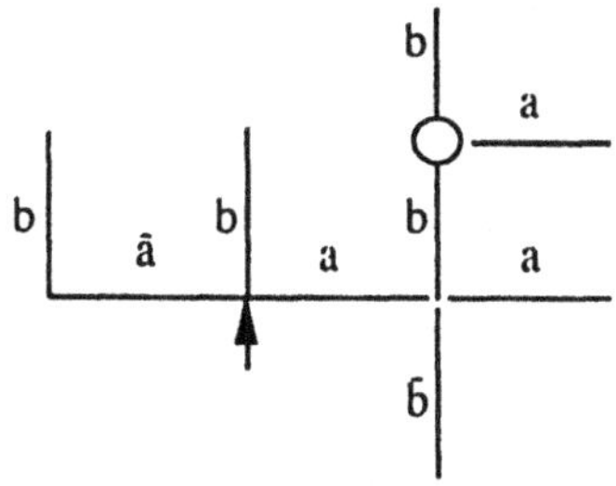

Figure 2: the tree associated with the word $\bar{a}b\bar{b}ab\bar{b}a\bar{b}ba\bar{a}ba\bar{a}b\bar{b}$

This characterization shows that the restriction to $R(\Sigma)$ of the application β is injective. We will identify $R(\Sigma)$ with its image by β, i.e., with the elements of $Inv(\Sigma)$

that are associated with a tree reduced to a walk with no turn, i.e., with a *path*. Since this facilitates greatly the notations we state it explicitely:

Convention: we identify the three subsets $R(\Sigma)$, $\alpha(R(\Sigma))=F(\Sigma)$ and $\beta(R(\Sigma)) \subseteq Inv(\Sigma)$. Thus, e.g., according to the context, $a\bar{b}\bar{b}\bar{a}b$ represents either a word over the four letter alphabet $\{a,\bar{a},b,\bar{b}\}$ or its image $\alpha(a\bar{b}\bar{b}\bar{a}b)$ in the free group generated by $\{a,b\}$ or its image $\beta(a\bar{b}\bar{b}\bar{a}b)$ in the free inverse monoid whose associated birooted tree is:

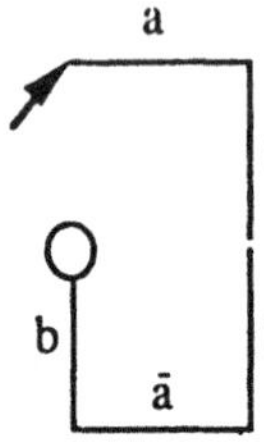

With a minor abuse of notation, for all $x\in Inv(\Sigma)$, we denote by $T(x)$ the common images $T(u)$ of all the possible representatives u of the class x. Then Theorem 3.1.1 can be rephrased as follows:

(3.1.3) *the application* $x \rightarrow (T(x), \pi(x))$ *is a bijection of the semigroup* $Inv(\Sigma)$ *into the set of all pairs consisting of a finite prefix-closed subset* $X \subseteq F(\Sigma)$ *and an element of* $F(\Sigma)$ *that belongs to* X.

A *reduced factorization* (where the monoid $Inv(\Sigma)$ is understood) is a product $u_1u_2...u_n$ where $u_1,u_2,...u_n$ are reduced non empty words and considered as a product of elements in the free group $F(\Sigma)$ no cancellation occurs between two consecutive factors.

Example 3.1.3: if $u_1 = b\bar{b}a$, $u_2=\bar{a}bb$ and $u_3=\bar{b}a$ then the products u_1u_2, u_2u_3 and u_3u_2 are not reduced but u_1u_3, u_2u_1 and u_3u_1 are. □

In terms of the previous characterization, the product of two elements x, $y\in Inv(\Sigma)$ behaves similarly to a semidirect product. Indeed, we have:

(3.1.4) $I(x)=(T(x),\pi(x))$ *and* $I(y)=(T(y),\pi(y))$ *implies*
$I(xy)=(T(x) \cup \pi(x)T(y), \pi(xy))$

As an immediate consequence, the idempotents are easily characterized:

(3.1.5) *an element* $x\in Inv(\Sigma)$ *is an idempotent if and only if* $\pi(x)=1$

Based on the right ideals they generate we can define a partial order on the elements of $Inv(\Sigma)$:

$x \geq y$ *if and only if there exists* z *such that* $xz=y$.

This ordering can be interpreted in terms of the Munn's characterization as can be readily verified:

(3.1.6) *given two elements* $x,y \in Inv(\Sigma)$ *we have* $x \geq y$ *if and only if* $T(x) \subseteq T(y)$

In view of (3.1.4) and (3.1.6) the following properties are more or less straightforward:

(3.1.7) *for all* $u \in R(\Sigma)$, *all idempotents* $e^2=e$ *and all* $y \in Inv(\Sigma)$ *we have:*
$u \geq ey$ *and* $u \not\geq e$ *implies* $u \geq y$

(3.1.8) *for all* $u, v \in R(\Sigma)$ *starting with different letters, and all* $x \in Inv(\Sigma)$ *we have:*
$v \geq ux$ *implies* $\bar{u}v \geq x$

(3.1.9) *for all reduced factorizations* uv *and all* $x \in Inv(\Sigma)$ *we have:*
$uv \geq ux$ *implies* $v \geq x$

(3.1.10) *for all idempotents* $e^2=e$ *we have:*
$$e = \prod_{u \in R(\Sigma)} \{ u\bar{u} \mid u \geq e;\ u \in R(\Sigma) \}$$

As a consequence of (3.1.10) we get a useful test for verifying that two elements are equal:

Proposition 3.1.2 *Given* $x,y \in Inv(\Sigma)$ *we have* $x=y$ *if and only if the following two conditions hold:*

(i) $\pi(x)=\pi(y)$

(ii) *for all* $u \in R(\Sigma)$ *we have:* $u \geq x$ *if and only if* $u \geq y$

3.2 Standard decompositions in free inverse monoids

The main purpose of this section is to cope with the non cancellativity of the monoid $Inv(\Sigma)$ by defining a standard decomposition of each element. Consider an element x in $Inv(\Sigma)$ such that $\pi(x)=u_1u_2...u_n$ where the factorization is reduced. The word $u_1u_2...u_n$ is the label of a unique path starting from the root of the tree $T=T(x)$. The notion of standard decomposition relatively to the factorization $u_1u_2...u_n$ can be defined in terms of a decomposition of T into subtrees $T_0,T_1,...,T_n$ as follows. If $n=0$ then $T_0=T$. Otherwise, let $T^{(1)}$ be the subtree of T that hangs from the node labelled by u_1. Then T_0 is the tree obtained by deleting $T^{(1)}$ from T. The process is then recursively applied to the subtree $T^{(1)}$ along with the factorization $u_2...u_n$.

Example 3.2.1: the standard decomposition of the element illustrated in Figure 3 relatively to the factorization $u_1u_2u_3$ where u_1=aba, u_2=baa and $u_3=\bar{b}a$ is pictured in Figure 4.

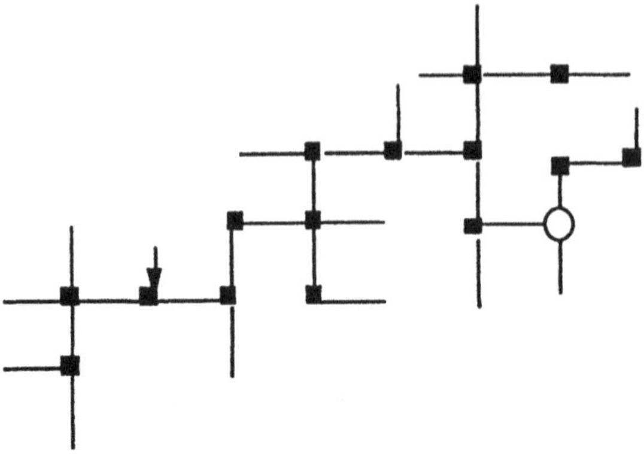

Figure 3: the original tree T

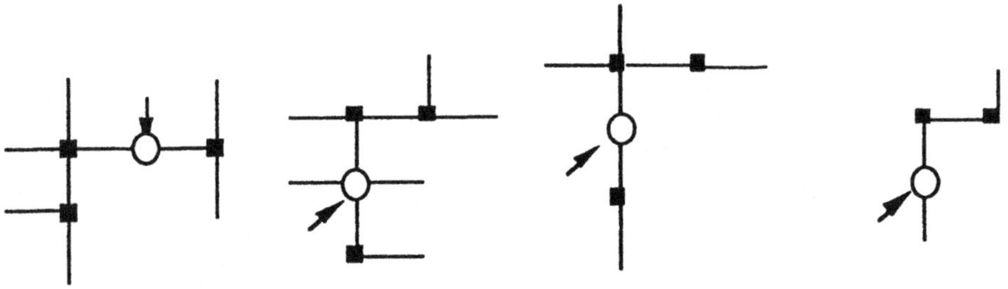

Figure 4: the standard decomposition of T into T_0, T_1, T_2, T_3

Before giving the formal definition we start with a technical Lemma:

Lemma 3.2.1 *Let* $x \in \mathrm{Inv}(\Sigma)$ *and let* $\pi(x)=uv$ *be a reduced factorization. Assume we have* $u=u'c$ *where* $c \in \Sigma$ *is the last letter of* u. *Then there exist a unique idempotent* $e^2=e$ *and a unique element* $y \in \mathrm{Inv}(\Sigma)$ *such that the following holds*:

(i) $x=euy$

(ii) $u' \geq e$ *and* $u \not\geq e$

(iii) $\bar{c} \not\geq y$

Proof Unicity: assume e and y satisfy (i), (ii) and (iii). We have:

$$e= \prod_{w \in R(\Sigma)} \{ w\bar{w} \mid w \geq x \text{ and u is not a prefix of w} \}$$

Indeed, if u is a prefix of w then $w \geq e$ implies $u \geq e$, contradiction. Conversely, assume that u is not a prefix of w and that $w \not\geq e$ holds. Then by (3.1.7) we have $w \geq uy$. By (3.1.8) $\bar{c} \not\geq y$ implies that u is a prefix of w, contradiction. This shows that e is unique. By the same properties, y is unique since $\pi(y)=v$ holds and $T(y)=T_u(x)$, where:

$$T_u(x) =\{ w \in R(\Sigma) \mid uw \text{ is reduced and } uw \in T(x) \}$$

Existence: consider the idempotent e such that:

$$T(e)=T(x)-uT_u(x)$$

and let $y\in Inv(\Sigma)$ be the element such that:

$$T(y)=T_u(x) \text{ and } \pi(y)=v$$

Then by (3.1.4) $x=euy$ holds and (ii) and (iii) are satisfied.□

We now turn to the formal definition of a standard decomposition. Let $u_1u_2...u_n$ be a reduced factorization. For $i=1,...,n$, set $u_i=u_i'c_i$ where c_i is the last letter of u_i. A *standard decomposition relatively to* $u_1u_2...u_n$ is a product:

$$e_0u_1e_1u_2...u_ne_n$$

where $e_0,e_1,...,e_n$ are idempotents satisfying the conditions:

(i) for all $i=1,...,n$, $u_i \not\geq e_{i-1}$ and $u_i' \geq e_{i-1}$

(ii) for all $i=1,...,n$, $\bar{c}_i \not\geq e_i$

The main result here states that all elements of $Inv(\Sigma)$ have at most one standard decomposition relatively to some reduced factorization:

Proposition 3.2.2 *Let* $x\in Inv(\Sigma)$ *and let* $u_1u_2...u_n$ *be a reduced factorization of* $\pi(x)$. *Then there exists a unique standard decomposition such that:*

$$x=e_0u_1e_1u_2...u_ne_n \qquad (3.2.1)$$

Proof By induction on n. If $n=1$ then this is Lemma 3.2.1. Assume $n>1$ and apply the same Lemma with $u=u_1$ and $v=u_2...u_n$: $x=e_0u_1y$. Because of the induction hypothesis y has a unique decomposition: $y=e_1u_2...u_ne_n$. Since $\bar{c}_1 \not\geq y$ holds we have $\bar{c}_1 \not\geq e_1$ which shows that the decomposition (3.2.1) is standard. The unicity follows from the unicity of the factorization of Lemma 3.2.1 and, via the induction hypothesis, from the unicity of the standard decomposition of y. □

We refer to (3.2.1) as to the *standard decomposition of* x *relatively to* $u_1u_2...u_n$.

4. CONJUGACY

We first investigate the relation of conjugacy on idempotents by showing that idempotents form a single class of conjugacy.

Proposition 4.1 *Assume* x *is an idempotent and* $xz=zy$ *holds for some* y. *Then* y *is an idempotent. Conversely, for all idempotents* x,y *there exists* z *such that* $xz=zy$ *holds.*

Proof Clearly, if $I(x)=(T(x),\pi(1))$, $I(y)=(T(y),\pi(y))$ and $I(z)=(T(z),\pi(z))$ then we have $\pi(z)=\pi(zy)$, i.e., $\pi(y)=\pi(1)$ and y is an idempotent by (3.1.5). Conversely, if x and y are two idempotents, then $xxy=xy=xyy$.

As a consequence of the above we observe that the restriction of conjugacy to idempotents is not the closure of the relation of transposition since, e.g., $[a\bar{a}]_{trans}=\{a\bar{a}\}$. If x is not an idempotent the situation is quite different since we will show that the relation of conjugacy is the transitive closure of the relation of transposition. But we first verify that we have an infinite hierarchy.

Proposition 4.2 *For all* $k>0$ *we have*: $\mathrm{Trans}^{k-1} \subset \mathrm{Trans}^{k}$

Proof Because of (3.1.4), each element x is transposed with only finitely many different y. If for some integer $k \geq 0$, $\mathrm{Trans}^{k-1} = \mathrm{Trans}^{k}$ were true then there would be only finitely many elements y such that $\mathrm{Trans}^*(x,y)$ would hold. However, consider $x=a$ where $a \in \Sigma$ is a letter. We define the following element for all $i,j \geq 0$:

$$x_{i,j} = \bar{a}^i a^{i+j+1} \bar{a}^j$$

Clearly, the result will be established if we verify that $\mathrm{Trans}^{i+j}(x, x_{i,j})$ holds for all $i,j \geq 0$. This is achieved by induction on $i+j$. Indeed, using the factorization:

$$x_{i,j} = \bar{a}^i a^{i+j+1} \bar{a}^j = \bar{a}^i a^i\, a\bar{a}a\, a^j \bar{a}^j = \bar{a}^i a^i\, a^{j+1}\, \bar{a}^{j+1} a \;= a^{j+1}\, \bar{a}^{j+1}\, \bar{a}^i a^i\, a\bar{a}a$$
$$= (a^{j+1}\, \bar{a}^{j+1})(a\, \bar{a}^{i+1} a^{i+1})$$

and transposing: $(a\, \bar{a}^{i+1} a^{i+1})(a^{j+1}\, \bar{a}^{j+1}) = \bar{a}^i a^i\, a\bar{a}a\, a^{j+1}\, \bar{a}^{j+1} = \bar{a}^i a^{i+j+2}\, \bar{a}^{j+1} = x_{i,j+1}$

we get $\mathrm{Trans}^{i+j+1}(x, x_{i,j+1})$.

Similarly, the previous computation leads to the following factorization:

$$x_{i,j} = \bar{a}^i a^{i+j+1} \bar{a}^j = (a^{j+1}\, \bar{a}^{j+1})(a\, \bar{a}^{i+1} a^{i+1}) = (a^{j+1}\, \bar{a}^j)(\, \bar{a}^{i+1} a^{i+1})$$

By transposition $(\, \bar{a}^{i+1} a^{i+1})\, (a^{j+1}\, \bar{a}^j) = \bar{a}^{i+1} a^{i+j+2}\, \bar{a}^j = x_{i+1,j}$

we get $\mathrm{Trans}^{i+j+1}(x, x_{i+1,j})$ which completes the verification. □

Our main theorem establishes the fact that restricted to non idempotents, the relation of conjugacy is the transitive closure of the relation of transposition.

Theorem 4.3 *For all pairs of elements* $x,y \in \mathrm{Inv}(\Sigma)$ *that are not idempotents we have*:

(4.1) $\qquad xz=zy$ *for some* $z \in \mathrm{Inv}(\Sigma)$

if and only if there exist an integer $n \geq 0$ *and a sequence* $x=x_0, x_1 \ldots, x_n=y$ *of elements in* $\mathrm{Inv}(\Sigma)$ *such that for* $i=0,\ldots n-1$ *there exist* u,v *satisfying* $x_i=uv$ *and* $x_{i+1}=vu$.

Proof Let us make a general observation. Assume we want to prove the result for x and y and there are elements x',y' such that $\mathrm{Trans}^*(x,x')$ and $\mathrm{Trans}^*(y,y')$ hold. Because of the inclusion $\mathrm{Trans}^* \subseteq \mathrm{Conj}$ we obtain $x'z'=z'y'$ for some z'. It then suffices to prove the result for x' and y'. We will apply this remark in an equivalent way by imposing successive conditions on x and y.

It is helpful to assume a linear order is given on Δ which allows to define the alphabetic ordering on Δ^* (cf. 2.1).

Claim 1: *given* $x \in \mathrm{Inv}(\Sigma)$ *there exists* x' *such that* $\mathrm{Trans}(x,x')$ *holds and* $\pi(x')$ *is cyclically reduced.*

Indeed, if $\pi(x)$ is not cyclically reduced then we have $\pi(x)=\bar{t}vt$ where v is cyclically reduced. Let:

$$e_0 \bar{t} e_1 v e_2 t e_3$$

be the standard decomposition of x relatively to $\bar{t}vt$.

Then the element $x'=ve_2te_3e_0\bar{t}e_1$ satisfies Trans(x,x') and $\pi(x')=v$ is cyclically reduced.

Claim 2: *let* $x\in \mathrm{Inv}(\Sigma)$ *such that* $\pi(x)$ *is cyclically reduced. Then there exists* x' *such that* Trans(x,x') *holds and* $\pi(x')$ *is the conjugate of* $\pi(x)$ *(as element of* Δ^**) that is minimal in the alphabetic ordering.*

Indeed, let $\pi(x)=v_1v_2$ where v_2v_1 is minimal in the alphabetic ordering. Consider the standard decompostion of x relatively to v_1v_2:

$$e_0v_1e_1v_2e_2$$

Then the element $x'=e_1v_2e_2e_0v_1$ satisfies Trans(x,x') and $\pi(x')=v_2v_1$. □

Claim 3: *let* $x\in \mathrm{Inv}(\Sigma)$ *such that* $\pi(x)=u^n$ *for some integer* $n>0$ *and some cyclically reduced element* u. *Set*:

$$p=\max\{k\geq 0 \mid \bar{u}^k \geq x\} \text{ and } q=\max\{k\geq 0 \mid u^{n+k} \geq x\}.$$

and consider the standard decomposition of u^pxu^q *relatively to* u^{n+p+q} :

$$e_{-p}ue_{-p+1}u...ue_{-1}ue_0ue_1u...ue_nue_{n+1}u...ue_{n+q}.$$

Define $y=uf_1u...uf_n$ *where for* $i=1,...,n$ *we have*:

$$f_i=\prod_{-p\leq j\leq n+q,\ j-i\equiv o \bmod n} e_j$$

Then $\mathrm{Trans}^*(x,y)$ *holds.*

By induction on p+q. If p+q=0, i.e., $x=e_0ue_1u...ue_n$, then Trans(x,y) holds where $y=ue_1u...ue_ne_0$. Because of $\bar{\bar{x}}=x$ we may assume $q>0$. Observe that $x=\bar{u}^pu^pxu^q\bar{u}^q$. Indeed, if $g_0\pi(x)g_1$ is the standard decomposition of x relatively to $\pi(x)$, then by (3.1.9) we have $u^q\geq g_1$, thus by (3.1.10) $u^q\bar{u}^q\geq g_1$ and $g_1u^q\bar{u}^q=g_1$. Similarly, we have $\bar{u}^p\geq g_0$, thus $\bar{u}^pu^p\geq g_0$ and finally $\bar{u}^pu^pg_0=g_0\bar{u}^pu^p=g_0$ which yields:

$$\bar{u}^pu^pxu^q\bar{u}^q=\bar{u}^pu^px=x$$

Using the standard decomposition of u^pxu^q relatively to u^{n+p+q} as above, we obtain:

$$x=te_0ue_1u...ue_nue_{n+1}u...ue_{n+q}\bar{u}^q$$

where

(4.2) $t=\bar{u}^pe_{-p}ue_{-p+1}u...ue_{-1}u$

is an idempotent. Consider the element y:

$$y=ue_{n+1}u...ue_{n+q}\bar{u}^qte_0ue_1u...ue_n$$
$$=te_0ue_{n+1}u...ue_{n+q}\bar{u}^q\ ue_1ue_2...ue_n$$

Then x and y are transposed. We have:

$$e_{n+1}u...ue_{n+q}\bar{u}^q\ u=e_{n+1}u...ue_{n+q}\bar{u}^{q-1}\bar{u}u$$
$$=e_{n+1}u\bar{u}ue_{n+2}...ue_{n+q}\bar{u}^{q-1}=e_{n+1}ue_{n+2}...ue_{n+q}\bar{u}^{q-1}$$

and since this element is an idempotent, we get:

$$y=te_0ue_1e_{n+1}u...ue_{n+q}\bar{u}^{q-1}ue_2...ue_n.$$

Similarly:

$$e_{n+2}u...ue_{n+q}\bar{u}^{q-1}u=e_{n+1}u...ue_{n+q}\bar{u}^{q-2}$$

is an idempotent and we have:

$$y=te_0ue_1e_{n+1}ue_2e_{n+2}u...ue_{n+q}\bar{u}^{q-2}ue_3...ue_n.$$

and so forth. There are two possibilities:

case 1: if $q \geq n$ then we end up in:

(4.3) $y=e_{-p}ue_{-p+1}u...ue_{-1}ue_0ue_1e_{n+1}ue_2e_{n+2}u...ue_ne_{2n}ue_{2n+1}ue_{2n+2}...ue_{n+q}\bar{u}^{q-n}$

Setting $q'=q-n$, we have:

$\max\{k \geq 0 \mid \bar{u}^k \geq y\}=p$ and $\max\{k \geq 0 \mid u^{n+k} \geq y\}=q'$.

and furthermore (4.3) is the standard decomposition of y relatively to the factorization $u^p y u^{q'}$.

case 2: if $q<n$ then we end up in:

(4.4) $y=e_{-p}ue_{-p+1}u...ue_{-1}ue_0ue_1e_{n+1}ue_2e_{n+2}u...ue_qe_{n+q}ue_{q+1}...ue_n$

We have:

$\max\{k \geq 0 \mid \bar{u}^k \geq y\}=p$ and $\max\{k \geq 0 \mid u^{n+k} \geq y\}=0$.

and furthermore (4.3) is the standard decomposition of y relatively to the factorization $u^p y$.

In both cases we may conclude by induction □

Because of the previous claims we may assume that if (4.1) holds then $\pi(x)=\pi(y)$ is cyclically reduced and minimal in its conjugacy class. Futhermore, we get: $\pi(x)\pi(z)=\pi(z)\pi(y)$ which implies that $\pi(x)=\pi(y)=u^n$ and $\pi(z)=u^p$ for some reduced element u and some $n,p>0$. Moreover, we may assume that the two standard decompositions of x and y relatively to u^n are $x=e_0ue_1u...ue_n$ and $y=g_0ug_1u...ug_n$ respectively and that they satisfy:

$$(4.5) \quad \begin{cases} e_0u=g_0u=u \\ u \not\geq e_n \text{ and } u \not\geq g_n \end{cases}$$

Consider the standard decomposition of z relatively to u^p:

$$z=f_0uf_1u...uf_p$$

Claim 4: $\bar{c} \not\geq f_0$ where c is the last letter of u and $u \not\geq f_p$.

Indeed, let $\bar{u}^k\bar{u}'$ be the largest prefix v of a power of $\bar{u}$ satisfying $v \geq f_0$ and assume it is different from the empty word. Because of the equality:

$$xz=e_0ue_1u...ue_nf_0uf_1u...uf_p=f_0uf_1u...uf_pg_0ug_1u...ug_n=zy$$

we have, via (3.1.7) and (3.1.8), $\bar{u}^k\bar{u}' \geq xz=e_0ue_1u...ue_nf_0uf_1u...uf_p$, i.e., $\bar{u}^{k+n}\bar{u}' \geq f_0$ contradicting the maximality of k.

Now we turn to the second inequality $u \not\geq f_p$ and assume that for some maximum $k>0$ we have $u^k \geq f_p$. Then $u^{k+n+p} \geq xz=zy$ holds and we get:

$\max\{r \mid u^r \geq zx=f_0uf_1u...uf_pg_0ug_1u...ug_n\}=\max\{p+k,p+n\}<k+n+p$

contradicting the maximality of k. □

As a result of claim 4, $e_0ue_1u...ue_nf_0uf_1u...uf_p$ and $f_0uf_1u...uf_pg_0ug_1u...ug_n$ are standard decompositions of the same element $xz=zy$, so by the unicity we may identify the idempotents that are in the same positions. Because of $e_0=f_0$, we get the further simplification:

$$e_0ue_1u...ue_nuf_1u...uf_p = f_0uf_1u...uf_pug_1u...ug_n$$

which implies that the two sequences:

$$e_1,...,e_n,f_1,...,f_p$$

and

$$f_1,...,f_p,g_1,...,g_n$$

be equal. This is precisely the equation of conjugacy in free monoids, implying the existence of an integer $0<k<n$ such that: $e_1=g_{k+1}$, $e_{n-k}=g_n$, $e_{n-k+1}=g_1$, ..., $e_n=g_k$ completing the proof. □

Now, we turn to the decidability aspect of the problem of conjugacy in free inverse monoids and we show, somehow surprisingly, that it has the lowest possible complexity since it is linear. We suppose that the two elements of the free inverse monoid are given by two representatives in Δ^*.

Theorem 4.4 *Given two words* $x,y\in\Delta^*$, *it can be tested in time* $O(m)$, *where* $m=\max\{|x|,|y|\}$ *whether or not* $xz \underset{E}{\approx} zy$ *holds for some word* $z\in\Delta^*$.

Proof The proof consists essentially in evaluating the complexity of the different operations carried out in the theorem 4.3. Since this latter proof uses standard decompositions, we need the following technical result that evaluates the complexity of computing standard decompositions:

Lemma 4.5 *Let* $x\in\Delta^*$ *be an arbitrary word and assume* $\alpha(x)=u_1u_2...u_n$ *where the product is reduced. Then the standard decomposition relatively to the factorization* $u_1u_2...u_n$ *can be computed in time* $O(|x|)$.

Proof Clearly, it suffices to determine the idempotents $e_0,e_1,...,e_n$ such that $\beta(x)=e_0u_1e_1u_2...u_ne_n$. This can be done in one pass while scanning the word from left to right. For all prefixes y of x there is a unique reduced factorization:

(4.6) $\alpha(y)=u_1...u_pz$

where $p\geq 0$, u_{p+1} is not a prefix of z. The idea is to maintain the values p and z while scanning the word.

More technically, the data structures that we use are:

i) an array Idem that contains the idempotents $e_0,e_1,...,e_n$
ii) an array Fact that stores the factors $u_1,u_2,...,u_n$
iii) the variable z and the integer $0\leq p\leq n$ as in (4.6)

Reading the word from left to right we determine a walk in the tree that represents the element $\beta(x)$. Keeping track of p and z determines the subtree of the idempotent we find ourselves in at a given moment. This leads to the following procedure in pseudo-Pascal:

```
p ← 0; z ← 1;                           {initialization of
for i ← 0 to n do Idem[i]=1;             the variables}

while x≠1 do begin          {x is the suffix still to be read}
      c ← First(x);         {reads the leftmost letter of x}
```

```
x ← LeftDelete(x):    {deletes the leftmost letter of x}
d ← Last(Fact[p]);    {d is the rightmost letter of the factor u_p}
case "compare z with 1 and u_{p+1}" of
1 {z=1 and dc≠1}: ( if Fact[p+1]=c then begin
                                         {u_{p+1} is reduced to c}
                       p ← p+1; z ← 1 end
                       else begin
                       Idem[p] ← Concat(Idem[p],c); z ← c end )

2 { z=1 and dc=1}: ( if Fact[p]=d then z ← 1      {u_p is reduced to d}
                       else
                       z ← RightDelete(Fact[p]); {z is obtained from u_p
                                  by deleting its rightmost letter}
                       p ← p-1 )

3 {z≠1 and zc≠Fact[p+1]} ( z ← Reduce(z,c);
                       Idem[p] ← Concat(Idem[p],c); )

4 { z≠1 and zc=Fact[p+1]} ( z ← 1; p ← p+1)

endcase
endwhile;
```

The functions First (returns the first letter of a word), Last (returns the last letter of a word), LeftDelete (returns the word after deletion of its leftmost letter), RightDelete (returns the word after deletion of its rightmost letter), Reduce (reduces, if necessary the product of a word with a letter, cf. the function ρ of section 3.1), Concat (concatenates in Δ^*, a word with a letter) can easily be implemented in constant time. We can also perform the test zc=Fact[p+1] in constant time provided 1) we keep track of the integer $|z| - |t|$ where t is the longest common prefix of z and u_{p+1} and 2) for any $0 \leq j \leq |u_{p+1}|$, we can have access to the j-th letter of u_{p+1} in constant time. □

We now turn to the proof of the theorem. This is done by evaluating the complexity of each step of the proof of Theorem 4.3. We first establish that starting with x (resp. y), it requires $O(m)$ to compute the element x' (resp. y') that satisfies claims 1 and 2 of theorem 4.2. Indeed, consider claim 1. The factorization $\pi(x)=\bar{t}vt$ where v is cyclically reduced, can be computed in linear time. Then Lemma 4.5 guarantees that the decomposition $x=e_0\bar{t}e_1ve_2te_3$ and thus the word $e_1ve_2te_3e_0\bar{t}$ is also computed in linear time. Similarly, consider claim 2. The factorization $\pi(x)=v_1v_2$ where v_2v_1 is minimal in the lexicographical order can be achieved in linear time (cf., [Bo] or [Du]). Again by Lemma 4.5 the decomposition $x=e_0v_1e_1v_2e_2$ and thus the element $e_1v_2e_2e_0v_1$ can be computed in linear time.

As seen in the proof of Theorem 4.3, we eventually have to compare two sequences of idempotents. In order to achieve linear time, we need to define some kind of canonical representation a idempotents. This is done by assigning to each $x \in \Delta^*$, a word that encodes the traversal of the tree T(x) formed by the vertices visited in a prefix order (cf., e.g., [AHU], p. 83). Thus, given an idempotent we first show how to determine the tree of visited vertices then we associate with this tree a word of the language of Dyck primes (cf.,e.g., [Be], p. 63).

More precisely, the set of all vertices of the Cayley graph $\Gamma(\Sigma)$ visited by x can be viewed as a tree labelled by the set $\Delta \cup \{1\}$: the root is the unique vertex labelled by 1 and the label of a vertex different from the root is equal to the last letter of the reduced word encoding the path from the root to the vertex. Clearly, the root has at most one child labelled by a given letter in Δ. Furthermore, each vertex labelled by some $a \in \Delta$ has at most one child labelled by some letter in $\Delta - \{\bar{a}\}$. The procedure goes as follows where the variable T denotes $T(x)$ and CurrentNode denotes the vertex currently visited:

```
T is reduced to the root labelled by 1;
CurrentNode ← 1;
while x≠1 do begin
    a ← First(x);
    if CurrentNode does not end with ā then
    create, if necessary, a child labelled by a whose father is CurrentNode;
    CurrentNode ← Reduce(CurrentNode,a);
    x ← LeftDelete(x)
end;
```

Now, we show how to encode any subtree T of $\Gamma(\Sigma)$ rooted at the node $v \in R(\Sigma)$, by some word $w(T)$ of the language of Dyck primes. We augment the alphabet Δ with two new symbols # and $\bar{\#}$ (where $\bar{\bar{\#}}=\#$) and enumerate the elements of Δ: $a_1, a_2, \ldots, a_{2n}$. Then we define $w(T)$ recursively as follows:

i) if T is reduced to its root then $w(T)=\# \, \bar{\#}$

ii) if $v=1$, i.e., if T is rooted at the root of $\Gamma(\Sigma)$, then:

$$w(T)=\# \, w(T_1) \, w(T_2) \ldots w(T_{2n}) \, \bar{\#}$$

where T_i, $i=1,\ldots 2n$ is the word associated with the subtree hanging from the root of T and labelled by $a_i \in \Delta$

iii) if u ends with $a \in \Delta$ and $\bar{a}=a_i \in \Delta$, then:

$$w(T)=\# \, w(T_1) \ldots w(T_{i-1}) \, w(T_{i+1}) \ldots w(T_{2n}) \, \bar{\#}$$

where T_j, $j=1,\ldots 2n$ is the word associated with the subtree hanging from the root of T and labelled by $a_j \in \Delta$.

It is clear from the definition that $w(T)$ can be determined in time linear with respect to the size of the tree.

So far if $\pi(x') \neq \pi(y')$ then the original words x and y are certainly not representatives of conjugate elements in $Inv(\Sigma)$. Thus we may assume $\pi(x')=\pi(y')$ and compute its root u in linear time: $\pi(x')=\pi(y')=u^n$. By Lemma 4.5 we may obtain in linear time the standard decompositions of x' and y' relatively to the factorization u^n. By claim 4, x' and y' lead to the factorizations $e_0ue_1u...ue_n$ and $g_0ug_1u...ug_n$ such that x' and y', i.e., x and y, are conjugate if and only if the sequence $e_1,e_2,...,e_n$ is cyclically equal to the sequence $g_1,g_2,...,g_n$. Let \$ be a new symbol and consider the words:

(4.7) $T(e_1)\$T(e_2)\$...\$T(e_n)\$$ and $T(g_1)\$T(g_2)\$...\$T(g_n)\$$

Because of the extra symbol \$, the sequences $e_1,e_2,...,e_n$ and $g_1,g_2,...,g_n$ are cyclically equal if and only if so are the two words in (4.7). One more final use of the above quoted algorithm in free monoids completes the proof. □

Acknowledgment I would like to thank J.P. Duval for having suggested an improvement in the original evaluation of the the complexity of the conjugacy problem.

REFERENCES

[Be] Berstel J., "Transductions and Context-Free languages", Teubner Verlag, 1979.

[Bo] Booth K., S., Lexicographically least circular substrings, Inform. Proc. Let., 4, (1980), pp. 240-242

[Du] Duval J.-P., Factorizing words over an ordered alphabet, Journal of Algorithms, 4, (1983), pp. 363-381

[Dub] Duboc C., On some equations in free partially commutative monoids, Theoret. Comput.Sci., 46, 1986, pp. 159-174

[Ho] Howie J.M., "An Introduction to Semigroup Theory", Academic Press, (1976)

[La] Lallement G., "Semigroups and Combinatorial Applications", J. Wiley, New York, (1979).

[LS] Lentin A. & M.P. Schützenberger, A combinatorial problem in the theory of free monoids, Proc. University of North Carolina, (1967), 128-144

[LyS] Lyndon R. & Schützenberger M.-P., The equation $a^M=b^Nc^P$ in a free group, Mich. Math. J., 9, (1962), 289-298

[Me] Meakin J., Automata and the word problem, Proc. of the 16-th LITP Spring School on Theoretical Computer Science, Lecture Notes in Computer Science, 386, pp. 89-103

[MKS] Magnus W., Karrass A. and Solitar D., *Combinatorial Group Theory*, 2nd ed., Dover, New York, (1976)

[Mu] Munn W.D., Free Inverse Semigroups, Proc. London Math. Soc., 3, 1974, pp. 385-404

[NO] Narendram P. and Otto F., The problems of cyclic equality and conjugacy for finite complete rewriting systems, 47, (1986), pp. 27-38

[NOW] Narendram P., Otto F. and Winklmann, The Uniform Conjugacy Problems for Finite Church-Rosser Thue Systems Is NP-Complete, 63, (1984), pp. 58-66

[Os] Osipova V.A., On the conjugacy problem in semigroups, Proc. Steklov Inst. Math., 133, (1973), pp. 169-182

[Ot] Otto F., Conjugacy in monoids with a special Church-Rosser presentation is Decidable, Semigroup Forum, 29, (1984), pp. 223-240

[S1] Scheiblich H.E., Free Inverse Semigroups, Semigroup Forum, 4, 1972, pp. 351-359

[S2] Scheiblich H.E., Free Inverse Semigroups, Proc. Amer. Math. Soc., 8, 1973, pp. 1-7

[St] Stephen J.B., The word problem for inverse monoids and related questions, to appear

[Zh] Zhang L., Conjugacy in special Monoids, to appear in J. of Algebra

General A- and AX-Unification via Optimized Combination Procedures

Franz Baader[1] and Klaus U. Schulz[2]

[1] DFKI, Stuhlsatzenhausweg 3, 6600 Saarbrücken 11, Germany
[2] CIS, University Munich, Leopoldstr. 139, 8000 München 40, Germany

Abstract. In a recent paper [BS91] we introduced a new unification algorithm for the combination of disjoint equational theories. Among other consequences we mentioned (1) that the algorithm provides us with a decision procedure for the solvability of general A- and AI-unification problems and (2) that Kapur and Narendran's result about the NP-decidability of the solvability of general AC- and ACI-unification problems (see [KN91]) may be obtained from our results. In [BS91] we did not give detailled proofs for these two consequences. In the present paper we will treat these problems in more detail. Moreover, we will use the two examples of general A- and AI-unification for a case study of possible optimizations of the basic combination procedure.

1 Introduction

For a long time, most applications of equational unification presupposed algorithms which solve the enumeration problem, i.e. the problem to enumerate in a compact way the set of all solutions of a given set of equations modulo an equational theory E. The recent development of constraint approaches to logic programming (see e.g., [JL87, Co90]), theorem proving (see e.g., [Bü90]) and term rewriting (see e.g., [KK89]) emphasizes the need for satisfiability checking and the decision problem received a new status. The decision problem for E is the problem to decide for a given set of equations whether they have a solution modulo E or not.

In the first part of this paper we shall study the decision problem in the context of two equational theories, namely the theories A and AI, expressing associativity respectively associativity plus idempotency of a function symbol. To be more precise, a distinction has to be made with respect to the signature which is used to built the terms of a unification problem.

In the simple case only the function symbol occurring in the equational axiom(s) is used, beside constants and variables. This type of problem will be called *unification with constants.* Associative unification problems with constants may be regarded as finite systems of equations in a free semigroup, often called word equations. Makanin's algorithm ([Ma77]) shows that the solvability of finite systems of word equations is decidable. Decidability of AI-unification problems with constants follows easily from the fact that idempotent semigroups are locally finite (see [Mc54]).

However, applications in constraint logic programming and other areas presuppose that terms may contain additional free function symbols. The Prolog III term

$$< FF(1, < 1 > \bullet x, 2) > \bullet x \bullet < FF(x, y, y), 1 >$$

for example may be regarded as a "generalized word" corresponding to a term

$$f(1, 1 \circ x, 2) \circ x \circ f(x, y, y) \circ 1$$

where "f" is a free function symbol, "1" and "2" are constants and "$\circ$" is an associative function symbol in infix notation. Constraints over lists as used in Prolog III stand in a close relationship with *general associative unification problems*, i.e., associative unification problems with free function symbols.

The decidability of general A- and AI-unification problems was open for a long time (compare Kapur and Narendran's table of known decidability and complexity results in unification theory [KN91]). A positive answer was obtained in our paper [BS91] where we used the fact that general E-unification may be regarded as an instance of the *combination problem*. The combination problem (see e.g, [BS91, Sc89]) is concerned with the question of how to derive unification algorithms (for the enumeration problem or the decision problem) for unification in the union of equational theories over disjoint signatures from unification algorithms in the single theories. In [BS91] the following general theorem was proved:

Theorem: *Let $E_1, \ldots, E_n$ be equational theories over disjoint signatures such that solvability of E_i-unification problems with linear constant restriction is decidable for $i = 1, \ldots, n$. Then unifiability is decidable for the combined theory $E_1 \cup \ldots \cup E_n$.*

Linear constant restrictions are induced by a linear order "$<$" on the set $X \cup C$ of variables and constants, demanding that, for a unifier θ, the constant c must not occur in $\theta(x)$ if $c > x$. The combination algorithm which is used to establish this result will be described in section 2. The description — which is the one given in [BS91] — is conceptually clear, but it is not the most appropriate one for efficient implementation since it does not try to minimize the number of possible choices in the non-deterministic steps. For this reason we will discuss optimized versions of the basic combination procedure in the context of general A- and AI-unification problems in section 3.

Let us return to the theorem. Obviously, every general A-unification (resp. general AI-unification) problem with set of free function symbols Ω may be regarded as a combination of the theory A (AI) with the free theory $F_\Omega = \{f(x_1, ..., x_n) = f(x_1, ..., x_n); f \in \Omega\}$. For the latter theory unification is simply unification in the empty theory. As a matter of fact, the solvability of Robinson unification problems with linear constant restrictions is decidable. If at least one unifier for such a problem satisfies the constant restrictions, then the most general unifier will satisfy them, and vice versa.

In order to prove that solvability of general *A*- (*AI*-) unification problems is decidable it remains to show that solvability of *A*- (*AI*-) unification problems with linear constant restrictions is decidable. These problems, and the related problems of *A*- (*AI*-) unification with *partially specified* linear constant restrictions (which arise from our optimization methods) will be discussed in sections 4 and 5.

Another problem which was solved only recently is the complexity of general *AC*- and *ACI*-unification, where *AC* (*ACI*) is the theory of one associative and commutative (and idempotent) function symbol. Kapur and Narendran [KN91] could prove that both problems are NP-decidable. In the last section we shall show that this observation follows easily from our combination result.

2 The combination algorithm

For the sake of convenience we shall restrict the presentation of the basic algorithm to the combination of two theories. The combination of more than two theories can be treated analogously. Before we can start with the description of the algorithm we have to introduce some notation.

Let E_1, E_2 be two equational theories built over disjoint signatures Ω_1, Ω_2, and let $E = E_1 \cup E_2$ denote their union. Since we are only interested in elementary E-unification[3], we can restrict our attention to terms built from variables and symbols of $\Omega_1 \cup \Omega_2$. The elements of Ω_1 will be called *1-symbols* and the elements of Ω_2 *2-symbols*. A term t is called *i-term* iff it is of the form $t = f(t_1, ..., t_n)$ for an i-symbol f $(i = 1, 2)$. A subterm s of a 1-term t is called *alien subterm* of t iff it is a 2-term such that every proper superterm of s in t is a 1-term. Alien subterms of 2-terms are defined analogously. An i-term s is *pure* iff it contains only i-symbols and variables. An equation $s \doteq t$ is pure iff there exists an $i, 1 \leq i \leq 2$, such that s and t are pure i-terms or variables; this equation is then called an i-equation. Please note that according to this definition equations of the form $x \doteq y$ where x and y are variables are both 1- and 2-equations. In the following, the symbols x, y, z, with or without indices, will always stand for variables.

Example 1. Let Ω_1 consist of the binary (infix) symbol "$\circ$" and Ω_2 of the unary symbol "h", let $E_1 := \{x \circ (y \circ z) = (x \circ y) \circ z\}$ be the theory which says that "$\circ$" is associative, and let $E_2 := \{h(x) = h(x)\}$ be the free theory for "h".

[3] i.e, unification where the terms of the unification problems are built over the signature of E.

The term $y \circ h(z \circ h(x))$ is a 1-term which has $h(z \circ h(x))$ as its only alien subterm. The equation $h(x_1) \circ x_2 \doteq y$ is not pure, but it can be replaced by two pure equations as follows. We replace the alien subterm $h(x_1)$ of $h(x_1) \circ x_2$ by a new variable z. This yields the pure equation $z \circ x_2 \doteq y$. In addition, we consider the new equation $z \doteq h(x_1)$. This process of replacing alien subterms by new variables is called *variable abstraction.* It will be the first of the five steps of our combination algorithm.

Combination Algorithm — Basic Form

The input for the *combination algorithm* is an elementary E-unification problem, i.e., a system $\Gamma_0 = \{s_1 \doteq t_1, \ldots, s_n \doteq t_n\}$, where the terms $s_1, \ldots, t_n$ are built from variables and the function symbols occurring in $\Omega_1 \cup \Omega_2$, the signature of $E = E_1 \cup E_2$. The first two steps of the algorithm are deterministic, i.e., they transform the given system of equations into one new system.

Step 1: variable abstraction. Alien subterms are successively replaced by new variables until all terms occurring in the system are pure. To be more precise, assume that $s \doteq t$ or $t \doteq s$ is an equation in the current system, and that s contains the alien subterm s_1. Let x be a variable not occurring in the current system, and let s' be the term obtained from s by replacing s_1 by x. Then the original equation is replaced by the two equations $s' \doteq t$ and $x \doteq s_1$. This process has to be iterated until all terms occurring in the system are pure. It is easy to see that this can be achieved after finitely many iterations. Now all the terms in the system are pure, but there may still exist non-pure equations, consisting of a 1-term on one side and a 2-term on the other side.

Step 2: split non-pure equations. Each non-pure equations of the form $s \doteq t$ is replaced by two equations $x \doteq s, x \doteq t$ where the x are always new variables.

It is quite obvious that these two steps do not change solvability of the system. The result is a system which consists of pure equations. The third and the fourth step are nondeterministic, i.e., a given system is transformed into finitely many new systems. Here the idea is that the original system is solvable iff at least one of the new systems is solvable.

Step 3: variable identification. Consider all possible partitions of the set of all variables occurring in the system. Each of these partitions yields one of the new systems as follows. The variables in each class of the partition are "identified" with each other by choosing an element of the class as representative, and replacing in the system all occurrences of variables of the class by this representative.

Step 4: choose ordering and theory indices. This step doesn't change a given system, it just adds some information which will be important in the next step. For a given system, consider all possible strict linear orderings $<$ on the variables of the system, and all mappings *ind* from the set of variables into the set of theory indices $\{1, 2\}$. Each pair $(<, ind)$ yields one of the new systems obtained from the given one.

The last step is again deterministic. It splits each of the systems already obtained into a pair of pure systems.

Step 5: split systems. A given system Γ is split into two systems Γ_1 and Γ_2 such that Γ_1 contains only 1-equations and Γ_2 only 2-equations. These systems can now be considered as unification problems with linear constant restriction. In the system Γ_i, the variables with index i are still treated as variables, but the variables with alien index $j \neq i$ are treated as free constants. The linear constant restriction for Γ_i is induced by the linear ordering chosen in the previous step.

The output of the algorithm is thus a finite set of pairs (Γ_1, Γ_2) where the first component Γ_1 is an E_1-unification problem with linear constant restriction, and the second component Γ_2 is an E_2-unification problem with linear constant restriction.

Proposition 1. *The input system Γ_0 is solvable if and only if there exists a pair (Γ_1, Γ_2) in the output set such that Γ_1 and Γ_2 are solvable.*

A proof of this proposition was given in [BS91]. It was also shown how the combination procedure can be used in order to generate a complete set of unifiers for the combined theory, provided that there are algorithms which enumerate complete sets of unifiers for unification problems with constant restrictions in the single theories.

As mentioned in the introduction, the presentation given above is conceptually simple but inefficient since no attempt was made to minimize the number of possible choices in the non-deterministic steps. Let us now give an optimized version of the combination algorithm in a particular context.

3 Optimization for Combination with the Free Theory

Before we treat the cases of general *A*- and *AI*- unification let us consider the situation where an arbitrary equational theory E is combined with an instance of the free theory

$$F_\Omega := \{f(x_1, \ldots, x_n) = f(x_1, \ldots, x_n); f \in \Omega\}$$

in order to obtain a unification algorithm for general E-unification.

We shall assume that the variables occurring in the unification problems are elements of a countable set Z which has fixed standard linear ordering $<_Z$. The following optimized combination algorithm rules out several choices which are possible in the non-deterministic steps of the non-optimized version.

Optimized Combination Algorithm — Version 1

Context: Combination of an equational theory E with the free theory F_Ω. Decision problem for general E-unification.

The input is a finite system Γ_0 of equations between terms built from variables, free constants and function symbols belonging to Ω or the signature of E. Without loss of generality we assume that Γ_0 does not contain an equation of the form $x \doteq y$ or $x \doteq a$ where x, y are variables and a is a free constant. In the following procedure, Γ_i denotes the system which is reached after step i. If all equations of Γ_i are pure, then $\Gamma_{i,F}$ denotes the subsystem of all equations which only contain free function symbols, $\Gamma_{i,E}$ denotes the subsystem with the equations containing function symbols from the signature of E. The first three steps are deterministic:

Step 1: variable abstraction. as before.
Step 2: split non-pure equations. as before.
Step 3: first variable identification. We solve the free subsystem $\Gamma_{2,F}$, treating all variables as variables and using standard unification in the free theory. If $\Gamma_{2,F}$ is unsolvable, then we stop with failure. In the following we shall assume that $\Gamma_{2,F}$ is solvable, with most general unifier (mgu) τ_1, say. Let Y_2 be the set of variables occurring in Γ_2. The maximal subsets of variables with same image under τ_1 define a partition π_1 of Y_2. Based on this partition we now identify variables in Γ_2, as described in step 3 of the non-optimized version. We obtain the system Γ_3 with set of variables Y_3.

The following steps are non-deterministic.

Step 4: second variable identification. We choose a partition π_2 of Y_3 and identify variables in Γ_3. We obtain the system Γ_4 with set of variables Y_4. If $\Gamma_{4,F}$ is unsolvable, or if the mgu τ_2 of $\Gamma_{4,F}$ identifies two distinct variables x and y of Y_4, then we stop this path with failure and backtrack. In the following we shall assume that $\Gamma_{4,F}$ is solvable with mgu τ_2 such that $\tau_2(x) \neq \tau_2(y)$ for all $x \neq y$ in Y_4.
Step 5: choose theory indices and ordering. Step 4 of the basic version is modified, imposing several restrictions on the indices *ind* and the linear orderings $<$ which may be selected. We first choose a variable indexing *ind* which satisfies the following condition: for every variable $x \in Y_4$,

- if $\tau_2(x) = f(t_1, \ldots, t_n)$ for a free function f, then $ind(x) = F$.

Let $Y_{4,F} \subseteq Y_4$ denote the set of all variables x with $ind(x) = F$, let $Y_{4,E} \subseteq Y_4$ be the remaining set of variables. We now choose a linear ordering $<_F$ on $Y_{4,F}$ which satisfies the following restriction: for all variables $x, x' \in Y_{4,F}$:

- if $\tau_2(x')$ is a subterm of $\tau_2(x)$, then $x' <_F x$.

Now $<_F$ will be extended to a linear ordering $<$ of Y_4. As a consequence of the following two conditions, this step is completely deterministic: for all $y, y_1, \ldots, y_k \in Y_{4,E}$,

- if y does not occur in any term $\tau_2(x)$, for $x \in Y_{4,F}$, then $x < y$ for all $x \in Y_{4,F}$. Otherwise, if x is the minimal element of the ordering $<_F$ such that y occurs in $\tau_2(x)$, then $y < x$. If x has an immediate predecessor x' in the ordering $<_F$, then $x' < y$,
- if $y_1 < y_2 < \ldots < y_k$ is a sequence of consecutive E-variables in $<$ (i.e., without F-variables in between), then $y_1 <_Z y_2 <_Z \ldots <_Z y_k$,

Eventually we check if the corresponding condition holds also for F-variables:

- if $x_1 < x_2 < \ldots < x_k$ is a sequence of consecutive F-variables in $<$, then $x_1 <_Z x_2 <_Z \ldots <_Z x_k$.

If the last condition is violated, then we stop with failure. In the other case the equational system is not modified, $\Gamma_5 := \Gamma_4$.

Step 6: split systems. as before. We obtain the systems $\Gamma_{6,F}$ and $\Gamma_{6,E}$ with linear constant restrictions induced by $<$.

Since, by construction of $<$, τ_2 is a solution of $\Gamma_{6,F}$, the systems of type $\Gamma_{6,E}$ may be regarded as the output of the combination procedure.

Proposition 2. *Γ_0 is solvable iff a system $\Gamma_{6,E}$ in the output set is solvable.*

Proof. We need a tool which was already introduced in [BS 91], namely unfailing completion. In the given context it suffices to apply this procedure to the theory E. This yields a possibly infinite ordered-rewriting system R which is confluent and terminating on ground terms. As in [BS 91], this system is also applied to terms containing variables from a fixed countable set of variables X_0 — for the completion these variables can simply be treated like constants. Let $T(\Omega, E, X_0)$ be the set of all terms built from function symbols in Ω, functions symbols of the signature of E and variables in X_0. Let $T_{\downarrow R}$ denote its R-irreducible elements. A substitution σ is called R-normalized on a finite set of variables Y iff $\sigma(y) \in T_{\downarrow R}$ for all variables $y \in Y$.

Since every output of the optimized version 1 is a possible output of the non-optimized procedure, correctness of the latter procedure implies correctness of the former.

Let us now prove completeness. Suppose that Γ_0 has a solution σ. We may assume that σ is also a solution of Γ_2 and that σ is R-normalized on the set Y_2 of variables occurring in Γ_2. In [BS 91] it is described how σ may be used to determine choices in the non-deterministic steps of the non-optimized version which lead to output systems $\Gamma_F^{(n.o.)}$ (free subsystem) and $\Gamma_E^{(n.o.)}$ (E-subsystem) which have solutions $\sigma_F^{(n.o.)}$ and $\sigma_E^{(n.o.)}$ which respect the linear constant restrictions. In more detail, the following choices were made:

(a) in step 3, partition $\pi^{(n.o.)}$ is chosen in such a way that variables x and y are identified iff $\sigma(x) = \sigma(y)$,

(b) in step 4 we define $ind^{(n.o.)}(x) := E$ iff $\sigma(x) = h(t_1, \ldots, t_n)$ where the function symbol h belongs to the signature of E, and $ind^{(n.o.)}(x) := F$ otherwise,

(c) the linear ordering $<^{(n.o.)}$ which is chosen is an arbitrary extension of the partial ordering $\prec$ defined by $y \prec x$ iff $\sigma(y)$ is a strict subterm of $\sigma(x)$.

We may also assume that $<^{(n.o.)}$ satisfies the following two conditions:

(d) if $x_1 <^{(n.o.)} x_2 <^{(n.o.)} \ldots <^{(n.o.)} x_k$ is a sequence of consecutive variables with $ind^{(n.o.)}(x_1) = ind^{(n.o.)}(x_2) = \ldots = ind^{(n.o.)}(x_k)$, then $x_1 <_Z x_2 <_Z \ldots <_Z x_k$.

In fact, permutation of elements in the subsequence $x_1 <^{(n.o.)} \ldots <^{(n.o.)} x_k$ does not modify the constant restrictions imposed on the output systems.

(e) if $ind^{(n.o.)}(x) = F$, $ind^{(n.o.)}(y) = E$ and if $y <^{(n.o.)} x$ is the immediate predecessor of x, then y occurs in $\sigma_F^{(n.o.)}(x)$.

Otherwise we could just change the order of these two variables, leaving all other order relations untouched. With this modification a new constant restriction for $\Gamma_F^{(n.o.)}$ is created, demanding that y must not occur in the value of x. But $\sigma_F^{(n.o.)}$ satisfies this condition and is still a solution of the modified system. For $\Gamma_E^{(n.o.)}$, one constant restriction is erased and $\sigma_E^{(n.o.)}$ is a solution.

We shall now show that some choices in the non-deterministic steps of the optimized version lead to the same output pair $(\Gamma_{6,F}, \Gamma_{6,E}) = (\Gamma_F^{(n.o.)}, \Gamma_E^{(n.o.)})$. Let us first show that step 3 will not fail. It is not difficult to see that $\Gamma_F^{(n.o.)}$ is just a restricted version of $\Gamma_{2,F}$: if we disregard constant restrictions, then $\Gamma_F^{(n.o.)}$ may be obtained from $\Gamma_{2,F}$ by identification of variables as described in (a), then treating E-variables as constants. Thus solvability of $\Gamma_F^{(n.o.)}$ implies solvability of $\Gamma_{2,F}$. The partition π_1 of step 3 is a refinement of $\pi^{(n.o.)}$: we claim that

$$(*)\ \tau_1(x) = \tau_1(y) \text{ implies } \sigma(x) =_E \sigma(y).$$

Since σ is R-normalized on Y_2 we then get $\sigma(x) = \sigma(y)$. Let us consider (*) in more detail, since the same kind of argument will also be used later. When we solve $\Gamma_{2,F}$ by a solved form algorithm (see e.g., [BB87, JK90]), then we add equations which are consequences of the equations in Γ_2. This follows from the fact that $=_E$ is a congruence relation and that we may derive the equations $r_1 = s_1, \ldots, r_n = s_n$ from an equation $f(r_1, \ldots, r_n) = f(s_1, \ldots, s_n)$ if f is a free function symbol. In particular every substitution which solves the equations in Γ_2 will also solve this derived equations. But the solved form contains the equations $x = \tau_1(x)$ and $y = \tau_1(y)$. Therefore $\sigma(x) =_E \sigma(\tau_1(x)) = \sigma(\tau_1(y)) =_E \sigma(y)$.

With the right choice in step 4 we will get the same variable identification as in the non-optimized procedure. This implies that σ is a solution of Γ_4. In particular σ solves $\Gamma_{4,F}$. By assumption (a), σ does not identify distinct variables of Y_4. It is easy to show that we may stop if $\tau_2(x) = \tau_2(y)$ for distinct variables $x, y \in Y_4$: with the same argument as above we see that $\sigma(x) =_E \sigma(\tau_2(x)) = \sigma(\tau_2(y)) =_E \sigma(y)$. Since σ is R-normalized it follows that $\sigma(x) = \sigma(y)$, a contradiction.

Let us now consider the restrictions of step 5. Note that $\Gamma_{4,F}$ and $\Gamma_F^{(n.o.)}$ are identical if we disregard constant restrictions in the latter system and treat E-variables as variables. If $\tau_2(x) = f(t_1, \ldots, t_n)$ for a free function symbol f, then $\sigma_F^{(n.o.)}(x)$ has top symbol f. Hence $\sigma_F^{(n.o.)}$ replaces x by a complex term. It does not treat it as a constant, therefore $ind^{(n.o.)}(x) = F$ and our choice in step 4 is conform. If $\tau_2(x')$ is a subterm of $\tau_2(x)$, then $\sigma(x') = \sigma(\tau_2(x'))$ is a strict subterm of $\sigma(\tau_2(x)) = \sigma(x)$. Thus $x' <^{(n.o.)} x$. It is now trivial to see that we may in fact choose $ind := ind^{(n.o.)}$ and "$<$" := "$<^{(n.o.)}$" in step 5. The rest is obvious. □

Remark. In version 1 of the optimized combination procedure some trivial hints were omitted which are relevant for efficient implementation. For example, the unifier τ_1 obtained in step 3 should be used when we decide solvability of $\Gamma_{4,F}$ in step 4. Another question which will not be discussed here is the question of how to determine an optimal order for the non-deterministic choices.

Partially Specified Linear Constant Restrictions

In some cases, one source of indeterminism in version 1 may be completely eliminated. The choice of a linear ordering in step 5 of version 1 of the optimized procedure can be avoided if it is possible to solve problems with ***partially specified linear constant restrictions*** for the theory E. Such a problem is given by an E-unification problem with constants, Γ, and a partial ordering $\leq$ on the variables and constants occurring in Γ. The problem has a solution iff there exists a linear extension of $\leq$ such that the unification problem with linear constant restriction induced by this extension has a solution.

Before we give a new version of the combination algorithm let us add a remark on the application area, in order to avoid misunderstandings. When we only ask for decidability per se, then unification problems with linear constant restrictions and unification problems with partially specified linear constant restrictions have the same status. If solvability of the former class of problems is decidable, then the same holds for the latter class of problems, simply because every partial ordering on a finite set has only a finite number of linear extensions. The interest in partially specified linear constant restrictions relies on the fact that for various equational theories E the effort to decide solvability of unification problems with partially specified linear constant restrictions does not exceed the effort to decide solvability of unification problems with linear constant restrictions. Examples are the theories A and AI, as will be shown. Thus, the "disjunctive" treatment of several linear orderings becomes superfluous and the corresponding indeterminism is really eliminated, not only postponed.

Optimized Combination Algorithm — Version 2

Context: Decision problem for general E-unification where decidability of E-unification problems with partially specified linear constant restrictions is decidable (in "non-disjunctive manner").

We simply proceed as in version 1 until we reach the point in step 5 where the variable indexing *ind* has been chosen. Now we do not choose a linear ordering, but propagate the partial ordering $\leq$ defined by

$$y \leq x \text{ iff } \tau_2(y) \text{ is a strict subterm of } \tau_2(x)$$

to the remaining E-subsystem, defining thus a system $\Gamma_{6,E}$ with partially specified linear constant restriction.

Proposition 3. *Γ_0 is solvable iff some system $\Gamma_{6,E}$ which is reached in version 2 is solvable.*

Proof. Suppose that $\Gamma_{6,E}$ is solvable, let $<$ be a linear extension of $\leq$ such that the corresponding unification problem with linear constant restriction $\Gamma'_{6,E}$ is solvable. The proof of proposition 2 shows that we may assume without loss of generality that $<$ is a possible choice in step 5 of version 1. Correctness of version 1 implies correctness of version 2.

Now consider the completeness proof for version 1. Since the linear ordering $<$ which was chosen for step 5 is a linear extension of the partial ordering $\leq$ which is used in version 2, completeness of version 2 follows immediately. □

Let us now consider the particular case of general unification problems in the theory

$$A := \{h(h(x,y),z) = h(x,h(y,z))\}$$

expressing associativity of the binary function symbol h.

Optimized Combination Algorithm — Version 3

Context: Decision problem for general A-unification.

We may assume without loss of generality that the input problem does not contain any equation of the form $f(s_1, ..., s_m) \doteq h(t_1, t_2)$ where f is a free function symbol. Since A is collapse-free we could immediately stop with failure in such a case.

Step 1 remains as in versions 1 and 2. Obviously Γ_1 cannot have non-pure equations. Thus step 2 may be omitted. Step 3 (first variable identification) remains almost as in versions 1 and 2. We add, however, a control step: if $\tau_1(x) = f(s_1, \ldots, s_n)$ for a free function symbol f and if $\Gamma_{3,A}$ contains and equation $x \doteq h(t_1, t_2)$, then we stop with failure. In step 4 (second variable identification) we do not identify variables x and y if $\tau_1(x) = f(t_1, \ldots, t_n)$ for a free function symbol f and if $\Gamma_{3,A}$ contains an equation $y \doteq h(s_1, s_2)$, or vice versa. Step 5 of version 2 is modified, eliminating some indeterminism in the choice of a variable indexing *ind*. We define $ind(y) = A$ whenever $\Gamma_{4,A}$ contains an equation $y \doteq h(s_1, s_2)$.

Correctness and completeness of version 3 follow immediately from the fact that the theory A is collapse-free. Note that we will get A-unification problems with partially specified linear constant restrictions as output of version 3. "Non-disjunctive" decidability of such problems is possible, as we shall sketch in the next section.

In the context of general AI-unification, where

$$AI := \{h(h(x, y), z) = h(x, h(y, z)), h(x, x) = x\},$$

we may again use version 2, as will be shown in section 5. But the additional modifications which were introduced for general A-unification cannot be used here since AI contains the collapse equation $h(x, x) = x$.

4 A-Unification with Linear Constant Restriction

We first want to prove the following

Theorem 4. *A-unification with linear constant restriction is decidable.*

Let us start with a simplified representation of A-unification problems with linear constant restriction. The associativity axiom $h(h(x, y), z) = h(x, h(y, z))$ allows to write every term built from variables, free constants and the associative symbol h in the normalform

$$h(t_1, h(t_2, \ldots, h(t_{n-1}, t_n) \ldots))$$

(where every t_i is either a variable or a free constant) which corresponds to rightmost-bracketing if infix notation is used. Obviously there is a bijective correspondence between terms of this form and words

$$t_1 t_2 \dots t_{n-1} t_n$$

where variables may be instantiated with non-empty words. This shows that every A-unification problem may be translated into a finite set of word equations, preserving unifiability in both directions. Every constant restriction of the A-unification problem translates directly into a constant restriction of the corresponding system of word equations. Thus we arrive at the following reformulation of our problem.

Theorem 5. *Unifiability of systems of word equations with linear constant restriction is decidable.*

There are two ways how we may prove this theorem. Let us first ignore matters of efficiency and just concentrate on the decidability result. In this case we may choose an even simpler formulation of the problem. A system (1) of word equations

$$U_1 \doteq V_1$$
$$U_2 \doteq V_2$$

may be rewritten into a single equivalent word equation (2)

$$U_1 a U_2 U_1 b U_2 \doteq V_1 a V_2 V_1 b V_2$$

where a and b are arbitrary distinct constants. Every unifier of (1) is a unifier of (2) and vice versa, thus both systems are also equivalent with respect to unifiability under linear constant restrictions. Therefore we may restrict considerations to single word equations.

Suppose now that a word equation WE is given with variables in $X = \{x_1, \dots, x_n\}$, constants in C and with linear constant restriction induced by a linear ordering $<$ on $X \cup C$. We want to decide whether WE has a unifier which satisfies the constant restriction.

For technical reasons we would like to use ground unifiers only. Let us consider an alphabet $C' = C \cup \{c\}$ where c is a constant not occurring in C for this purpose. Then WE has a unifier θ over $C \cup X$ which satisfies the linear constant restriction induced by $<$ if and only if WE has a solution (i.e, a ground unifier) θ' over C' which satisfies the linear constant restriction induced by "$<$": to obtain θ' from θ we just replace all occurrences of variables by the constant c. In the converse direction, all occurrences of c are replaced by the same variable.

We are now in a position to use the following general result from [Sh90]:

Theorem: *If WE is a word equation with variables $x_1, \ldots, x_n$ and constants in the alphabet C', and if $L_1, \ldots, L_n$ are regular languages over C', then it is decidable whether WE has a solution θ such that $\theta(x_i) \in L_i$ for $i = 1, \ldots, n$.*

Let $L_i := C_i^+$, where

$$C_i := \{a \in C; a < x_i\} \cup \{c\}$$

($i = 1, \ldots, n$). Obviously a solution θ of *WE* satisfies the linear constant restriction induced by $<$ iff $\theta(x_i) \in L_i$ for $i = 1, \ldots, n$. Thus the theorem implies that unifiability of word equations with linear constant restrictions is decidable, and thus — by disjunctive treatment — also with partially specified linear constant restrictions.

The algorithm which was used in [Sh90] in order to establish the theorem mentioned above is, however, more complicated than it would be necessary for the special purpose of regular languages of the form C_i^+. In connection with the initial translation of systems of word equations into a single word equation and with the disjunctive treatment of different linear orderings for general A-unification via version 1 of the combination procedure we get a rather inefficient decision procedure.

For the readers which are familiar with Makanin's algorithm let us sketch how a better solution might look like. In particular we want to show that it is possible to avoid

- the translation of a system of word equations into a single word equation, which doubles the data size,
- the disjunctive treatment of linear extensions of the partial ordering $\leq$ obtained in step 5 of the optimized version 2.

Thus, the prerequisite which is needed for version 3 is in fact available in the case of A-unification.

First, it is not difficult to show that Makanin's algorithm for deciding the solvability of words equations may directly be used for *systems* of word equations (see e.g, [Sh90]). Second, by means of simple additional devices it is possible to modify Makanin's algorithm in such a way that we keep control over all constants which will occur in the value of a variable, following a certain path of the search tree. For this purpose, a sequence of columns of the actual generalized equation is used to represent an actual value $V(x)$ of a variable x occurring in *WE*. If a subcolumn of a column of $V(x)$ contains a constant base of type $a \in C$, then the constant a will occur in $\theta(x)$ for every unifier θ of *WE* which corresponds to a successful node which may be found below the actual node of the search tree. After any transformation step, the values (sequences) of all variables have to be updated.

We may now run the algorithm with the partial order $\leq$ used as a "filter". This means that we stop a branch of the search tree as soon as the constants which occur in the (columns which represent the) actual values $V(x)$ of the variables x of *WE* show that the relation $\leq_V$ contains a cycle, where

$$y \leq_V x \text{ iff } y \leq x \text{ or } y \text{ occurs in } V(x).$$

A detailled description of this algorithm for word equations with partially specified linear constant restrictions would be rather long.[4] But most constructions are straightforward, starting from a modern description of Makanin's algorithm (e.g., [Sh91]).

5 AI-Unification with Linear Constant Restriction

Let h be a binary function symbol, and let *AI* denote the theory of idempotent semigroups, i.e.,

$$AI := \{h(h(x,y),z) = h(x,h(y,z)), h(x,x) = x\}.$$

As mentioned in the introduction, decidability of *AI*-unification problems with constants is an easy consequence of the fact that idempotent semigroups are locally finite. In fact, McLean has shown as early as 1954 (see [Mc54]) that finitely generated free idempotent semigroups are finite, and he gave a formula which can be used to calculate the cardinality c_n of the free idempotent semigroup in n generators:

$$c_n = \sum_{r=1}^{n} \left(\binom{n}{r} \prod_{i=1}^{r} (r-i+1)^{2^i} \right).$$

The method used to show this finiteness result can also be employed to get a decision procedure for the word problem for free idempotent semigroups (see e.g. [BC71]).

Putting these facts together one can conclude that, for a given finite set C of constant symbols (the generators), one can compute a finite set of representatives for the $=_{AI}$ classes of all terms built over h and these constants. In addition, for two such representatives s, t one can effectively find the representative of the class of $h(s,t)$. Since the theory *AI* is regular, all the elements of an *AI*-class contain the same constants as its representative.

Theorem 6. *AI-unification with linear constant restriction is decidable.*

[4] In order to prove termination of this algorithm, a straightforward generalization of Bulitko's theorem has to be proved.

Proof. Assume that Γ is an *AI*-unification problem with constants, and that $<$ is a linear ordering on the constants and variables occurring in Γ. If Γ does not contain any constant, then there are no constant restrictions and we have an elementary *AI*-unification problem. Thus we may assume that Γ contains at least one constant. If we ignore the constant restriction induced by $<$, then it is easy to see that Γ has an *AI*-unifier iff it has an *AI*-unifier such that the images of variables occurring in Γ contain only constants from Γ. In fact, one can substitute all other constants and variables occurring in the image of a unifier by constants from Γ, and thus ends up with a unifier satisfying the above condition. This is no longer possible if one has to fulfill a constant restriction. However, if we allow for one new constant (i.e., a constant not occurring in Γ) in the image, then we may substitute all constants beside the ones from Γ and all variables occurring in the image of a solution of the problem with constant restriction by this new constant, and end up with a unifier satisfying the constant restriction.

This shows that it is enough to look for a solution in the free idempotent semigroup in finitely many generators, namely the constants occurring in Γ plus one additional new constant. As mentioned above, this semigroup is effectively given by a finite set of representatives. Thus what we can do is try all possibilities of replacing the variables in Γ by these representatives, under the restriction imposed by the linear ordering. Depending on whether we find a solution this way, the problem with linear constant restriction is solvable or not. □

As mentioned in section 3, the number of nondeterministic choices in the combination procedure can be reduced if it is possible to solve problems with partially specified linear constant restrictions. Recall that such a problem is given by a unification problem with constants, Γ, and a partial ordering $\leq$ on the variables and constants occurring in Γ. This problem has a solution iff there exists a linear extension of $\leq$ such that the unification problem with linear constant restriction induced by this extension has a solution.

It is easy to modify the procedure described in the proof of Theorem 6 to cope with such problems. In fact, when choosing a representative s to be substituted for a given variable x, this tells us that x has to be made larger than all constants from Γ occurring in s. If these additional relationships are consistent with the current partial ordering, we augment the partial ordering with them. Otherwise, we can discard this choice.

Corollary 7. *AI-Unification with partially specified linear constant restriction is decidable in "non-disjunctive" manner.*

It should however be noted that the decision methods we have just described are only practicable if the number of constants occurring in Γ is very small. This is so because the cardinalities c_n of finitely generated free idempotent semigroups are growing very fast.

6 NP-Decidability of general AC- and ACI-unification

In this section we shall prove NP-decidability of general AC- and ACI-unification problems. The result is an immediate consequence of the following two claims:

Claim 1: *For a given input problem Σ of size n_0,*[5] *the (non-optimized) combination algorithm may be organized in such a way that*

- *(1) the size n_{i+1} of the system obtained after step $i+1$ is polynomial in the size n_i of the system created at the previous step ($i = 0, \ldots, 4$).*
- *(2) the non-deterministic choices of steps 3 and 4 may be computed by means of a polynomial number of non-deterministic choices between two possibilities,*
- *(3) the complexity of the deterministic computations in step $i+1$ is polynomial in n_i ($i = 0, \ldots, 4$).*

Claim 2: *Solvability of AC and ACI-problems with linear constant restrictions may be decided by NP-algorithms.*

Proof of claim 1:

Ad (1). In step 1, note that one replacement of an alien subterms introduces two occurrences of a new variable, but does not introduce any new subterm. The number of such steps does not exceed the number of (occurrences of) subterms in the input symbol, thus it does not exceed n_0. Thus $n_1 \leq 3n_0$. Obviously $n_2 \leq 2n_1$. In the remaining steps, the size of the system(s) does not grow.

Ad (2). In step 3, the number of variables is $n \leq n_2$. Suppose that we have variables $\{x_1, ..., x_n\}$. In order to compute a partition of this set we shall successively create a partition P_i of $\{x_1, ..., x_i\}$ for $i = 1, ..., n$. Of course $P_1 := \{\{x_1\}\}$. Suppose that some $P_i = \{p_{i,1}, ..., p_{i,k}\}$ has been computed for an $i < n$. Obviously $k \leq i$. To compute P_{i+1} we may either add x_{i+1} to $p_{i,j}$ and stop, or we may continue, considering $p_{i,j+1}$, (starting with $j = 1$). If we reach the point where $p_{i,k}$ is considered, we may either add x_{i+1} to $p_{i,k}$ or we create a new equivalence class $p_{i+1,k+1} := \{x_{i+1}\}$. Obviously the final partition P_n is reached after not more than n_2^2 nondeterministic choices between two alternatives. It is also clear that every partition of $\{x_1, ..., x_n\}$ is a possible outcome of the procedure.

Suppose that the variables $\{x_1, ..., x_m\}$ are left in the input system for step 4. First we choose nondeterministically a variable which is subtracted from the

[5] The size of a unification problem is the number n of occurrences of symbols in the terms of the equations occurring in the problem.

set. This takes at most m non-deterministic decisions between two possibilities. The variable which is chosen is minimal with respect to the linear order to be constructed. The next variable is chosen in the remaining system. By iteration, the linear order is constructed in less than m^2 non-deterministic choices between two possibilities. Indexing of variables takes other m non-deterministic steps with two possibilities.

Ad (3). This is easy for step 1 since every term is only treated once, replacing all alien subterms. The remaining steps are obvious.

Proof of claim 2:

AC-Unification with Linear Constant Restriction. It is a well-known fact that solving *AC*-unification problems with constants can be reduced to solving systems of linear equations over the nonnegative integers (see e.g., [St81, Fa84]). As an easy consequence one can show that solvability of *AC*-unification problems with linear constant restriction can be expressed as an integer programming problem, thus establishing NP-decidability. Instead of giving a formal presentation of this reduction, we shall illustrate it by an example.

Let h be a binary *AC*-symbol, x, y be variables, and c, d be constants. We consider the *AC*-unification problem with constants

$$\begin{aligned}\Gamma = \{&h(x,h(x,h(c,h(c,c)))) \doteq h(y,h(y,h(y,h(y,d)))),\\ &h(x,h(x,h(x,h(y,y)))) \doteq h(x,c)\},\end{aligned}$$

and the constant restriction induced by $c < x < d < y$. As mentioned before, it is enough to look for solutions which introduce the constants c, d occurring in Γ and one additional constant, say e. For each of these three constants, we introduce a system of linear equations. The variables occurring in these equations stand for the number of occurrences of the respective constant in the image of x and y, respectively, of possible solutions of Γ. The coefficients of these variables in the equations are the number of occurrences of x and y, respectively, in Γ. Thus we get the three systems

$$\begin{pmatrix}2&0\\3&2\end{pmatrix}\begin{pmatrix}x_c\\y_c\end{pmatrix}+\begin{pmatrix}3\\0\end{pmatrix}=\begin{pmatrix}0&4\\1&0\end{pmatrix}\begin{pmatrix}x_c\\y_c\end{pmatrix}+\begin{pmatrix}0\\1\end{pmatrix}$$

$$\begin{pmatrix}2&0\\3&2\end{pmatrix}\begin{pmatrix}x_d\\y_d\end{pmatrix}=\begin{pmatrix}0&4\\1&0\end{pmatrix}\begin{pmatrix}x_d\\y_d\end{pmatrix}+\begin{pmatrix}1\\0\end{pmatrix}$$

$$\begin{pmatrix}2&0\\3&2\end{pmatrix}\begin{pmatrix}x_e\\y_e\end{pmatrix}=\begin{pmatrix}0&4\\1&0\end{pmatrix}\begin{pmatrix}x_e\\y_e\end{pmatrix}$$

In addition, since we do not have a unit element for h, the variables x, y have to be substituted by nonempty terms. This is expressed by the inequalities

$$x_c + x_d + x_e > 0 \quad\text{and}\quad y_c + y_d + y_e > 0.$$

The *AC*-unification problem with constants, Γ, has a solution iff the above systems of equations have nonegative integer solutions satisfying the restriction imposed by these inequalities.

Now it should be obvious how to express the constant restriction with the help of some additional equations: If a constant must not occur in the image of a variable, the corresponding variable in the system of linear equations has to be zero. In our example, we get the additional equation

$$x_d = 0$$

because $x < d$ means that d must not occur in the image of x.

ACI-Unification with Linear Constant Restriction. Kapur and Narendran [KN91] have shown that solvability of *ACI*-unification problems with constants can be decided by a (deterministic) polynomial algorithm. This is done by transforming the conditions for solvability of such a unification problem into a set of propositional Horn clauses. This set of Horn clauses is satisfiable iff the original problem was unifiable. Since the size of this set of clauses is quadratic in the size of the unification problem, the fact that satisfiability of propositional Horn clauses is decidable in linear time (see [DG84]) shows that one ends up with a quadratic decision procedure for *ACI*-unifiability with constants.

Now we shall demonstrate that this result can easily be generalized to solvability of *ACI*-unification with constant restriction. To this purpose we briefly review Kapur and Narendran's transformation. Let Γ be an *ACI*-unification problem with constants, and let X and C be the variables and constants, respectively, occurring in Γ. Let C' be C augmented by one additional constant c_0. As explained in the proof of Theorem 6, it is enough to look for solutions introducing only these constants. For each pair $(x, c) \in X \times C'$, we take a propositional variable $P_{x,c}$. The intended semantics of this variable is that it is true iff c does not occur in the image of x for the substitution under consideration.

Now consider an equation $s \doteq t$. For all constants a occurring in s but not in t, one introduces a Horn clause

$$\bigwedge_{x \in V(t)} P_{x,a} \Longrightarrow \text{false},$$

where $V(t)$ denotes the variables occurring in t. The obvious meaning of this clause is that, in order to get a solution, a must be introduced by some variable of t, since it already occurs in s. The analogous formulae are built for the constants occurring in t but not in s. For the additional constant c_0, we have to add formulae saying that if this constant is not introduced on one side of the equation, it must not be introduced on the other side. Thus we have for all variables y occurring in s the formula

$$\bigwedge_{x \in V(t)} P_{x,c_0} \Longrightarrow P_{y,c_0}.$$

Of course, one also must take the analogous formulae where the role of s and t are exchanged. Finally, the fact that all variables $x \in X$ must be replaced by a nonempty term is expressed by the formulae

$$\bigwedge_{c \in C'} P_{x,c} \Longrightarrow \text{false}.$$

If we take these formulae for all the equations in Γ then we have obtained a set of Horn clauses which is satisfiable iff Γ has a unifier.

Obviously, this encoding is very convenient for expressing constant restrictions. The fact that c must not occur in the image of x can simply be expressed by the fact

$$\text{true} \Longrightarrow P_{x,c}.$$

To sum up, we have thus shown that solvability of *ACI*-unification problems with linear constant restriction can be decided by a (deterministic) quadratic algorithm.

Conclusion

We have studied general unification problems for the equational theories *A*, *AI*, *AC* and *ACI*. With the combination algorithm which was introduced in [BS91] it is possible to reduce general unification problems to Robinson unification plus unification problems with linear constant restrictions in the single theories. The latter class of problems was shown to be decidable for the four theories. From the proof for the theories *AC* and *ACI* and from the structure of the combination algorithm we obtained the result that solvability of general *AC*- and *ACI*-unification problems may be decided with NP-algorithms. For the theories *A* and *AI* we showed how to use the information obtained from the solution of the free subsystem which is separated by the combination algorithm in order to optimize this procedure, eliminating possible choices in the non-deterministic steps. The notion of a unification problem with partially specified linear constant restriction arose from our optimization technique and we demonstrated that this class of problems is decidable in non-disjunctive manner for the theories *A* and *AI*.

References

[BB87] K.H. Bläsius, H.-J. Bürckert, "Deduktionssysteme," Oldenbourg Verlag, München Wien (1987).

[BS91] F. Baader, K.U. Schulz, "Unification in the Union of Disjoint Equational Theories: Combining Decision Procedures," DFKI-Research Report RR-91-33, to appear in the *Proceedings of the 11th International Conference on Automated Deduction*, *LNCS* (1992).

[BC71] J.A. Brzozowski, K. Culik, A. Gabrielian, "Classification of Noncounting Events," *J. Computer and System Science* **5**, 1971.

[Bü90] H.-J. Bürckert, "A Resolution Principle for Clauses with Constraints," *Proceedings of the 10th International Conference on Automated Deduction, LNCS* **449**, 1990.

[Co90] A. Colmerauer, "An Introduction to PROLOG III," *C. ACM* **33**, 1990.

[DG84] W.F. Dowling, J. Gallier, "Linear Time Algorithms for Testing Satisfiability of Propositional Horn Formula," *J. Logic Programming* **3**, 1984.

[Fa84] F. Fages, "Associative-Commutative Unification," *Proceedings of the 7th International Conference on Automated Deduction, LNCS* **170**, 1984.

[JK90] J.P. Jouannaud, C. Kirchner, "Solving Equations in Abstract Algebras: A Rule-Based Survey of Unification," Preprint, 1990. To appear in the Festschrift to Alan Robinson's birthday.

[JL87] J. Jaffar, J.L. Lassez, "Constraint Logic Programming," *Proceedings of 14th POPL Conference*, Munich, 1987.

[KN91] D. Kapur, P. Narendran, "Complexity of Unification Problems with Associative-Commutative Operators," Preprint, 1991. To appear in *J. Automated Reasoning*.

[KK89] C. Kirchner, H. Kirchner, "Constrained Equational Reasoning," *Proceedings of SIGSAM 1989 International Symposium on Symbolic and Algebraic Computation*, ACM Press, 1989.

[Ma77] G.S. Makanin, "The Problem of Solvability of Equations in a Free Semigroup," *Mat. USSR Sbornik* **32**, 1977.

[Mc54] D. McLean, "Idempotent Semigroups," *Am. Math. Mon.* **61**, 1954.

[Sc89] M. Schmidt-Schauß, "Combination of Unification Algorithms," *J. Symbolic Computation* **8**, 1989.

[Sh90] K.U. Schulz, "Makanin's Algorithm - Two Improvements and a Generalization," *Proceedings of the First International Workshop on Word Equations and Related Topics IWWERT '90*, Tübingen 1990, Springer LNCS 572.

[Sh91] K.U. Schulz, "Word Unification and Transformation of Generalized Equations," CIS-Report 91-46, University of Munich, 1991 (see also this issue).

[St81] M. Stickel, "A Unification Algorithm for Associative-Commutative Functions," *J. ACM* **28**, 1981.

Word Equations With Two Variables

Witold Charatonik,
Institute of Computer Science
University of Wrocław
Wrocław, Poland

Leszek Pacholski
Institute of Mathematics
Polish Academy of Sciences
Wrocław, Poland

1 Introduction

The problem whether the set of all equations that are satisfiable in some free semigroup - or, equivalently, in an algebra of words with concatenation - is recursive (usually called the satisfiability problem for semigroup equations) was first formulated by A.A. Markov in early sixties (see [3]). Special cases of the problem were solved affirmatively by A.A. Markov (see [3]), Yu.I. Khmelevskiĭ [8], [7], G. Plotkin, [14] and A. Lentin [11]. The full positive solution, was given by G.S. Makanin in a paper [12], which is long and very technical.

Makanin's decision procedure for equational satisfiability in semigroups has received a lot of attention in the literature. Undoubtedly, this is because the notion of an algebra of words (or strings) with the operation of concatenation - is of fundamental importance in computer science: many algorithms and data structures refer to words. Thus, several improvements of Makanin's algorithm have been given by H. Abdulrab, J.-P. Pecuchet, K. Schulz, A. Kościelski and L. Pacholski (see [2], [13], [15], [9], and [10]), and attempts have even been made to implement the algorithm (see [1]). Moreover, related unification problems have been studied. In particular, J. Jaffar, in [6], basing on the Makanin's decision procedure, described an algorithm which, when an equation has a solution, generates all its solutions and halts if the set of solutions is finite.

An important fact used in the Makanin's algorithm and in the unification algorithms based on it, is that the periodicity exponent of a minimal solution of a word equation can be bounded by a recursive function of the length of the equation. In fact, V.K.Bulitko, in [4], proved that if d is the length of an equation, then the index of periodicity of its minimal solution (see below) does not exceed $(6d)^{2^{2^{d^4}}} + 2$. Kościelski and Pacholski ([9], [10]) forced this bound down to $2^{1.07d}$. They also prove a lower bound of $2^{0.29d}$ for the exponent of periodicity of minimal solutions of a word equation of length d.

Although the bound on the exponent of periodicity given by Kościelski and Pacholski gave an over-exponential improvement of the algorithm its complexity is still so high that it prohibits any applications in practice. Moreover, Kościelski and Pacholski [10] proved that the problem of the solvability of word equations is NP-hard, even if a linear bound is put on the length of possible solutions. Thus, for a given constant $c > 2$ the problem of the existence of a solution of length cd for an equation of length d is NP-complete. This implies, that there does not exist any fast algorithm, which decides solvability of all word equations and suggests that good algorithms can be found only for restricted classes of equations.

This paper contains the first report on our research project with the aim to describe classes of word equations for which fast algorithms, deciding solvability or giving actual solutions, exist. In this paper by "fast" we mean "deterministic polynomial time". Of course for many actual applications it would be better to consider more restricted classes like linear time or DTIME(nlog(n)). This problem will be considered in subsequent papers.

We consider equations which have at most two distinct variables. For such equations we give a deterministic polynomial time algorithm deciding their solvability. Our technique and algorithm is based on the notion of an "equation in exponent" which has been introduced by Yu.I. Khmelevskiĭ [7].

2 Preliminaria

The set of nonnegative integers is denoted by $\mathbb{N}$. For a finite set Σ, Σ^* is the set of words over Σ (the free semigroup generated by Σ), Σ^+ is the set of nonempty words over Σ, and Σ^c is the set of words of length c over Σ. ε denotes is the empty word, and $|W|$ denotes the length of a word W.

Let $\Sigma = \{a_1, \ldots, a_n\}$ and $\Xi = \{x, y\}$ be two disjoint alphabets, called respectively the alphabet of coefficients and the alphabet of variables. A word equation $\mathcal{E}$ over (Σ, Ξ) is a pair of words (W_1, W_2) (also denoted by $W_1 = W_2$), where $W_1, W_2 \in (\Sigma \cup \Xi)^+$. $|W_1 W_2|$ is the length of $\mathcal{E}$. A solution of $\mathcal{E}$ is a function $v : \Xi \to \Sigma^+$ such that $W_1(v(x)/x), v(y)/y)) = W_2(v(x)/x), v(y)/y))$, where $W(v(x)/x))$ denotes the word obtained from W by replacing each occurrence of x by $v(x)$. Given any function $v : \Xi \to \Sigma^*$, slightly abusing the notation, by the same letter v we denote the extension of v to the homomorphism $v : (\Sigma \cup \Xi)^* \to \Sigma^*$, which is the identity on Σ. Sometimes we identify the function v with the pair of words $(v(x), v(y))$. For words A_1, A_2 we write $A_1 < A_2$ if A_1 is a prefix of A_2. If $A_1, \ldots, A_n$ are words then $[A_i]_{i=1}^n$ denotes the concatenation of $A_1, \ldots, A_n$. A word A is primitive if $A \neq S^n$, for any word S and any integer $n > 1$. The length of a solution v is $|v(x)| + |v(y)|$. A solution is minimal if it has minimal length.

Below we shall give some preliminary results. Most of them can be found in [8].

Proposition 2.1 (Proposition 1.16 in [8]) *Let $A \in \Sigma^+$, $V, \Phi, \Psi \in \Sigma^*$, and assume that $V\Phi = A^a V \Psi$, for some $a > 0$. Then $V = A^t A_1$, for some $t \geq 0$ and $A_1 < A$.*

Notice that every equation with one variable is equivalent to one in the form

$$A[xA_i]_{i=1}^n = [xB_j]_{j=1}^m. \tag{1}$$

Lemma 2.2 (Proposition 1.19 in [8]) *If the equation (1) is solvable then it has a solution of length smaller then $M^2 + 3M$, where $M = \max_{i,j}\{n, m, |A_i|, |B_j|, |A|\}$.*

Corollary 2.3 *It can be decided in time $O(d^5)$ if a word equation $\mathcal{E}$ of length d with one variable has a solution.*

Proof. Without any loss of generality we can assume that $\mathcal{E}$ has the form $x\Phi = Px\Psi$, where $\Phi, \Psi \in (\Sigma \cup \{x\})^*, P \in \Sigma^+$. By Proposition 2.1 for any solution $v : \{x\} \to \Sigma^+$ of $\mathcal{E}$ we have $v(x) = P^t P_1$ for some integer t and a word $P_1 < P$. Moreover, by Proposition 2.2 $\mathcal{E}$ has a solution of length smaller than $d^2 + 3d$, so to decide if $\mathcal{E}$ is solvable it suffices to check if it is satisfied by one of $d^2 + 3d$ words of the form $P^t P_1$. This can be done in time $O(d^5)$ since there are $O(d^2)$ possibilities and in each case the word $v(x)\Phi Px\Psi(v(x)/x)$ has the length $O(d^3)$. ∎

Definition 2.4 *An exponential equation is an expression of the form $P_0[S_i^{\lambda_i} P_i]_{i=1}^n = Q_0[T_j^{\mu_j} Q_j]_{j=1}^m$, where λ_i, μ_j are integer variables, $P_i, Q_j \in \Sigma^*, S_i, T_j \in \Sigma^+$. A solution of such an equation assigns integer values to variables in such a way, that both sides of the equation become graphically identical.*

Lemma 2.5 (Proposition 2.4′ in [8]) *If an exponential equation $P_0[S^{\lambda_i} P_i]_{i=1}^n = Q_0[S^{\mu_j} Q_j]_{j=1}^m$ with two variables, (i.e. $\lambda_i, \mu_j \in \{\lambda, \mu\}$) is solvable, then it has a solution such that $\lambda \leq 4h^2 H$ or $\mu \leq 4h^2 H$, where $h = \max\{n, m, 8\}$, and $H = \frac{\max\{|S|,|P_i|,|Q_j|\}}{|S|}$*

Lemma 2.6 (Proposition 2.7 in [8]) *If an exponential equation $P_0[S^\lambda P_i]_{i=1}^n = Q_0[S^\lambda Q_j]_{j=1}^m$ with one variable λ is solvable, then it has a solution such that $\lambda \leq 2hH$, where $h = \max\{n, m, 8\}$, and $H = \frac{\max\{|S|,|P_i|,|Q_j|\}}{|S|}$*

Lemma 2.7 (Implicite in the proof of Proposition 2.8 in [8]) *If an exponential equation $[(S^\sigma C)^{\lambda_i} S^\sigma A_i]_{i=1}^n = [(S^\sigma C)^{\mu_j} S^\sigma B_j]_{j=1}^m$ with variables λ_i, μ_j, σ has a solution such that $\lambda_i, \mu_j \geq 3$, then it has a solution such that $\lambda_i, \mu_j \geq 3$ and $\sigma|S| + |C| \leq \max_{i,j}\{|A_i|, |B_j|\} + 2|SC|$.*

Corollary 2.8 *If an exponential equation $P_0[S^{\lambda_i} P_i]_{i=1}^n = Q_0[S^{\mu_j} Q_j]_{j=1}^m$, where $\lambda_i, \mu_j \in \{\lambda, \mu\}$ are integer variables, is solvable, then it has a solution such that $\lambda \leq 4d^3, \mu \leq 8d^5$ or $\mu \leq 4d^3, \lambda \leq 8d^5$, where $d = |P_0[SP_i]_{i=1}^n Q_0[SQ_j]_{j=1}^m|$, so it can be solved in time $O(d^{14})$.*

Proof. It is an easy consequence of Lemma 2.5 and Lemma 2.6. ∎

Definition 2.9 *A directed equation is an expression of the form $x\Phi \to y\Psi$ or $x\Phi \leftarrow y\Psi$, where $\Phi, \Psi \in (\Sigma \cup \Xi)^*$. A solution of $x\Phi \to y\Psi$ is any solution v of the equation $x\Phi = y\Psi$ such that $|v(x)| > |v(y)|$.*

Lemma 2.10 (Proposition 3.1 in [8]) *Given words $\Phi, \Psi \in (\Sigma \cup \Xi)^*$, and $B_j \in \Sigma^*$ for $j \leq b$, where b is an integer ≥ 1, let*

$$x\Phi \to [yB_j]_{j=1}^b x\Psi \tag{2}$$

be a directed equation with two variables x, y. For integers t, k and a word B such that $0 \leq t, 0 \leq k < b$, and $B < B_{k+1}$, we put

$$\sigma_1(x) = ([yB_j]_{j=1}^b)^t [yB_j]_{j=1}^k yB, \quad \sigma_2(x) = ([yB_j]_{j=1}^b)^t [yB_j]_{j=1}^k x,$$

and $\sigma_1(y) = \sigma_2(y) = y$.

Then if $(v(x), v(y))$ is a solution of (2) then exactly one of the two conditions below holds:

1. *For some k, t such that $0 \le k < b, 0 \le t$ and a prefix B of B_{k+1} we have*

$$v(x) = ([v(y)B_j]_{j=1}^{b})^t [v(y)B_j]_{j=1}^{k} v(y)B \tag{3}$$

and $v(y)$ is a solution of the equation

$$\sigma_1(x\Phi) = \sigma_1([yB_j]_{j=1}^{b} x\Psi) \tag{4}$$

obtained by applying σ_1 to (2).

2. *For some integers k, t such that $0 \le k < b, 0 \le k$ and a word X' we have*

$$v(x) = ([v(y)B_j]_{j=1}^{b})^t [v(y)B_j]_{j=1}^{k} X' \tag{5}$$

and $(X', v(y))$ is a solution of the equation

$$\sigma_2(x\Phi) = \sigma_2([yB_j]_{j=1}^{b} x\Psi) \tag{6}$$

obtained by applying σ_2 to (2).

Moreover, for any solution $v(y)$ of (4) the pair $(v(x), v(y))$, where $v(x)$ is defined by (3) is a solution of (2).

Finaly, for any solution $(X', v(y))$ of (6), the pair $(v(x), v(y))$, where $v(x)$ is defined by (5) is a solution of (2).

Lemma 2.11 (Proposition 4.4 in [8])*Let $\nu(\Phi)$ denote the pair (t_1, t_2), where t_1 (respectively t_2) is the number of occurrences of the variable x (respectively y) in the word Φ. Let $\bar{\Phi}$ denote the projection of the word Φ onto the alphabet Σ. Let $\Phi_1\Phi_2 = \Psi_1\Psi_2$ be an equation with two variables such that $\nu(\Phi_1) = \nu(\Psi_1)$ and $c = |\bar{\Phi}_1| - |\bar{\Psi}_1| \ge 0$. Then the following equivalence holds:*

$$\Phi_1\Phi_2 = \Psi_1\Psi_2 \iff \exists R \in \Sigma^c \; (\Phi_1 = \Psi_1 R \;\;\&\;\; R\Phi_2 = \Psi_2)$$

Lemma 2.12 (Proposition 5.10 in [8]) *The directed equation $xAy \to yBx$ is solvable if and only if there exist words $P, S \in \Sigma^*, Q \in \Sigma^+$ such that $A = PQS, B = SQP$.*

Definition 2.13 *For an equation*

$$xAy\Phi \to yBx\Psi \tag{7}$$

and a substitution σ such that $\sigma(x) = (xAyB)^t x$, $\sigma(y) = xAy$. equations $xAy\sigma(\Phi) \to yBx\sigma(\Psi)$ and $xAy\sigma(\Phi) \leftarrow yBx\sigma(\Psi)$ are called t-images of (7).

We say that equation (7) has the property α if it is equivalent to an equation of the form $xAyP\xi\Phi' \to yBxQ\eta\Psi'$, for some $P, Q \in \Sigma^*$, $P \ne Q$, $\Phi, \Psi \in (\Sigma \cup \Xi)^*$, $\xi, \eta \in \Xi$.

Lemma 2.14 (Proposition 7.4 in [8]) *If $A \ne B$, then either equation (7) is equivalent to the equation $xAy \to yBx$, or each t-image of (7) has the property α.*

Definition 2.15 *The exponent of periodicity of a word W is the maximal positive integer p such that $W = U_1U^pU_2$ for some words U_1, U_2 and a nonempty word U. The exponent of periodicity of a minimal solution of equation $\mathcal{E} = (W_1, W_2)$ is the greatest of the exponents of periodicity of the words $v(x), v(y)$, where v is a minimal (with respect to length) solution of $\mathcal{E}$.*

Lemma 2.16 *If p is the exponent of periodicity of a minimal solution of an equation of length $d \geq 6$ over $(\Sigma, \{x,y\})$, then $p \leq 4d^5$.*

A proof of this lemma is given in [5] following the idea of the proof of the main theorem of Chapter 3 in [9].

Lemma 2.17 *An equation $x\Phi = Px\Psi$, of length $d \geq 6$ with two variables x, y, and such that $P \in \Sigma^*$, $\Phi, \Psi \in (\Sigma \cup \{x,y\})^*$ is solvable if and only if it has solution v such that $v(x) = P^tP_1$ for an integer $t \leq 4d^5$ and a word $P_1 < P$.*

Proof. It is easy to see that any solution v of this equation is of the form $v(x) = P^tP_1$ for some integer $t \geq 0$ and $P_1 < P$. The bound $t \leq 4d^5$ follows from lemma 2.16. ∎

Definition 2.18 *For a word S and an integer M we write $\tau(S, M)$ if the following condition holds: there exist words A, A_1 and integer t such that $A_1 < A$, $|A| < M$ and $S = A^tA_1$.*

A reduction of an equation (WW_1, WW_2) consists of transformation of it into an equivalent equation (W_1, W_2), where words W_1, W_2 begin with different symbols. An equation is trivially unsolvable if after reduction its sides begin with different symbols from the alphabet of coefficients or exactly one of its sides is the empty word.

3 Basic equations

In this section we consider equations of the form $xAy\Phi = yBx\Psi$ of length d, with $A, B \in \Sigma^*$, $\Phi, \Psi \in (\Sigma \cup \Xi)^*$, and $|A| = |B|$. We distinguish two cases, $A = B$ and $A \neq B$.

3.1 The algorithm

Case 1 $A = B$

Make in parallel steps 1.1, 1.2, 1.3.

Step 1.1 For each suffix Z of A and prefixes Z_1, Z_2 of Z such that $Z_1A = Z^k$, and $Z_2A = Z^l$ for some $k, l \in I\!N$ substitute $Z^\lambda Z_1$ for x, $Z^\mu Z_2$ for y and solve the exponential equation with variables λ, μ obtained by this substitution.

Step 1.2 For each λ, μ such that either $(\lambda < 3, \mu \leq 4d^2)$ or $(\mu < 3, \lambda \leq 4d^2)$ or $(3 \leq \lambda \leq 4(12d^3)^3, 3 \leq \mu \leq (8(12d^3)^5)$ or $(3 \leq \mu \leq 4(12d^3)^3, 3 \leq \lambda \leq (8(12d^3)^5)$ substitute $(vA)^\lambda v$ for x, $(vA)^\mu v$ for y and solve the word equation with one variable v obtained by this substitution.

Step 1.3 Substitute $(vA)^\lambda v$ for x, $(vA)^\mu v$ for y and solve the exponential equation obtained in this way. This is an equation with two variables λ, μ, over alphabet $\Sigma \cup \{v\}$ of coefficients.

Case 2 $A \neq B$

Step 2.1 substitute y for x and solve the equation with one variable obtained by the substitution

Step 2.2 solve two systems of equations $(xAy \to yBx, \Phi = \Psi)$ and $(xAy \leftarrow yBx, \Phi = \Psi)$

Step 2.2.1 input: a system $xAy \to yBx, \Phi = \Psi$ of equations
output: at most one system of equations of the form $xAy \to yBx, P\xi\Phi' = \eta\Psi'$ $(\xi, \eta \in \Xi)$.

Use the following procedure to transform the equation $\Phi = \Psi$ to an equation of the form $P\xi\Phi' = \eta\Psi'$ with $P \neq \varepsilon$ (or to the empty or a trivially unsolvable equation):

REPEAT

- reduce the equation
- if the equation has the form $xA_1[\xi_i A_i]_{i=2}^n = (yB)^q xB_{q+1}[\eta_i B_i]_{i=q+2}^m$, then replace it with the equation $xA_1[\xi_i A_i]_{i=2}^n = x(Ay)^q B_{q+1}[\eta_i B_i]_{i=q+2}^m$
- if the equation has the form $xAyA_2[\xi_i A_i]_{i=3}^n = yB_1[\eta_i B_i]_{i=2}^m$, then replace it with the equation $yBxA_2[\xi_i A_i]_{i=3}^n = yB_1[\eta_i B_i]_{i=2}^m$

UNTIL none of the rules above applies.

If the equation obtained untill now is not of the form $P\xi\Phi' = \eta\Psi'$ then for each $t \leq d$ and $C < B$

- substitute $(yB)^t yC$ for x and solve the equation with one variable obtained in this way
- substitute $(xAyB)^t x$ for x, xAy for y, reduce the equation and if now it has the form $xAyBx\overline{\Phi} = yCxAy\overline{\Psi}$ then replace it with equation $yBxBx\overline{\Phi} = yCxAy\overline{\Psi}$ and reduce the last one

If this procedure gives the empty equation, then solve the equation $xAy \to yBx$ by finding decomposition $A = PQS, B = SQP$ with $Q \neq \varepsilon$. $x = QSQ, y = Q$ is a solution of this equation. If the procedure gives a trivially unsolvable equation, then give up this branch of algorithm.

Step 2.2.2 input: system of equations $xAy \to yBx$, $P\xi\Phi = \eta\Psi$ with $|P\xi\Phi\eta\Psi| = d$
For each $t \leq 4d^5$ and $P_1 < P$ substitute yx for x, and then $P^t P_1$ for y in the equation $xAyP\xi\Phi \to yBx\eta\Psi$ and solve equation with one variable obtained by this substitution.

3.2 Correctness of the algorithm

Theorem 3.1 (correctness of algorithm 3.1)

- *in case 1 an input equation is solvable if and only if one of the equation created in steps 1.1–1.3 has a solution*
- *in case 2 an input equation is solvable if and only if one of the equation (or systems of equations) created in steps 2.1–2.2 has a solution*
- *if output in step 2.2.1 is nonempty, then the input system is equivalent to the output system; otherwise the input system is solvable if and only if there exist a decomposition described in this step*
- *in step 2.2.2 the input system of equations is solvable if and only if one of the equations created in this step has a solution*

Proof. **Case 1.** If $A = B$, then $xAyA = yAxA$, so there exists such word $Z \in \Sigma^+$ that $xA = Z^k$ and $yA = Z^l$ for some $k, l \in I\!N$.

Step 1.1 corresponds to the case $|Z| \leq |A|$. If $|Z| > |A|$, then $Z = vA$ for some word v, so $x = (vA)^\lambda v$, and $y = (vA)^\mu v$ for some $\lambda, \mu \in I\!N$.

Step 1.2 corresponds to the case $\tau(v, M)$, where M is the maximal length of coefficient subword of Φ or Ψ. In this case $v = S^\sigma S_1$ for some $\sigma \in I\!N$, $S_1 < S$, and $|S| \leq M$, so $x = (S^\sigma S_1 A)^\lambda S^\sigma S_1$, and $y = (S^\sigma S_1 A)^\mu S^\sigma S_1$. If $\lambda < 3$ then (treating λ and σ as fixed) by proposition 2.7 of [8] we have $\mu \leq 4d^2$, and similarly for $\mu < 3$. If $\lambda, \mu \geq 3$ then by proposition 2.8 of [8] we have $\sigma \leq 6M$ and by substituting $(S^\sigma S_1 A)^\lambda S^\sigma S_1$ for x, and $(S^\sigma S_1 A)^\mu S^\sigma S_1$ for y we get an exponential equation of length less then $12d^3$ which can be solved using lemma 2.8.

Step 1.3 corresponds to the case $\neg\tau(v, M)$. After substituting $(vA)^\lambda v$ for x and $(vA)^\mu v$ for y (with fixed λ, μ) our equation gets the form $A_0[vA_i]_{i=1}^n = B_0[vB_j]_{j=1}^m$. Since we have $\neg\tau(v, M)$, proposition 1.16 of [8] gives us $n = m$, $A_i = B_i$ for $i \leq n$.

Case 2. Steps 2.1 and 2.2 correspond to the three possibilities: either $|x| = |y|$ (step 2.1), or $|x| < |y|$, or $|x| > |y|$ (step 2.2).

Step 2.2.1. Proof of correctness of this step can be found in the proof of propositions 7.4 and 5.10 of [8].

Step 2.2.2. From the definition of directed equation it follows that any solution of the system $xAy \to yBx, P\xi\Phi = \eta\Psi$ is such that $x = yx'$ for some nonempty word x'. After substitution yx for x the equation $P\xi\Phi = \eta\Psi$ gets the form $Py\Phi' = y\Psi'$ and the thesis follows from lemma 2.17. ∎

4 Reduction to basic equations

Now we consider an arbitrary equation of length d with two distinct variables. Reducing the same symbols in the beginning of both sides of the equation we can assume the equation has the form $P\xi\Phi = \eta\Psi$, where $P \in \Sigma^*, \xi, \eta \in \Xi = \{x, y\}$, and $\Phi, \Psi \in (\Sigma \cup \Xi)^*$

4.1 The algorithm

Step 1 input: an equation $P\xi\Phi = \eta\Psi$ of length d
output: $O(d)$ equations of the form $x\Phi' = y\Psi'$, each of length $O(d^2)$.

Case 1.1 $\xi = \eta$, and the equation has the form $Px\Phi = x\Psi$.
For each $t \leq 4d^5$ and each $P_1 < P$ substitute $P^t P_1$ for x and solve the equation with one variable obtained by this substitution.

Case 1.2 $\xi \neq \eta$, and the equation has the form $Px\Phi = y\Psi$,

Step 1.2.1 For each $P_1 < P$ substitute P_1 for y and solve the equation with one variable obtained by this substitution.

Step 1.2.2 Substitute Py for y and solve the equation $x\Phi' = y\Psi'$ obtained by this substitution.

Step 2 input: an equation $x\Phi = y\Psi$
output: two equations of the form $x\Phi \rightarrow y\Psi$

Step 2.1 Substitute y for x and solve the equation with one variable obtained by this substitution.

Step 2.2 Solve the directed equation $x\Phi \rightarrow y\Psi$

Step 2.3 Solve the directed equation $x\Phi \leftarrow y\Psi$

Step 3 input: an equation $x\Phi \rightarrow y\Psi$ of length d
output: $O(d^2)$ equations of length $O(d^2)$ of the form $xAy\Phi' = yBx\Psi'$

We can assume that input is of the form $xA\xi\Phi'' \rightarrow [yB_j]_{j=1}^{b} x\Psi''$ where $A, B_j \in \Sigma^*$ (if not, we consider the equivalent equation $x\Phi x \rightarrow y\Psi x$)

Step 3.1 For each $t \leq d$, $k < b$ and $B'_{k+1} < B_{k+1}$ substitute $([yB_j]_{j=1}^{b})^t [yB_j]_{j=1}^{k} yB'_{k+1}$ for x and solve the equation with one variable obtained by this substitution.

Step 3.2 For each $t \leq d$, $k \leq b$ substitute $([yB_j]_{j=1}^{b})^t [yB_j]_{j=1}^{k} x$ for x. In this way $O(d^2)$ equations of length $O(d^2)$ of the form $xAy\bar{\Phi} \leftarrow [yB'_j]_{j=1}^{b} x\bar{\Psi}$ are obtained.

Step 3.3 Substitute xy for y and get $Axy\bar{\Phi}(xy/y) = [yB'_j x]_{j=1}^{b} \bar{\Psi}(xy/y)$

Step 3.4 For each $A_1 < A$ substitute A_1 for y and solve the equation with one variable obtained by this substitution.

Step 3.5 Substitute Ay for y and get $xAy\Phi' = yB'_1 x\Psi'$

Step 4 input: an equation $xAy\Phi = yBx\Psi$ of length d
output: $O(d^2)$ equations of the form $xA'y\Phi' = yB'x\Psi'$ of length $O(d^2)$ with $|A'| = |B'|$

We can assume that $|A| - |B| = c > 0$ (if not then just replace A with B and x with y)

Step 4.1 For each $0 < i \leq c$ let A_i be the prefix of A of length $|B| + i$. For each $0 \leq j \leq |A_i|$ let A'_j be the suffix of A_i of length j. Let $R_{i,j}$ be the prefix of length c of the word $A'_j A_i$. Substitute $R_{i,j}$ for y and solve the equation with one variable obtained by this substitution.

Step 4.2 For each decomposition $A = PQS$ of the word A such that $P, S \in \Sigma^*, Q \in \Sigma^+$, let $R \in \Sigma^c$ be a word such that $RB = SQP$ (if such a word exists). Substitute yR for y and solve the equation $xAyR\Phi(yR/y) = yRBx\Psi(yR/y)$

4.2 The correctness

Theorem 4.1 (correctness of algorithm 4.1) *In each step of the algorithm 4.1 an input equation is solvable if and only if there exists a solvable equation created in this step.*

Proof. We will prove this theorem separately for each step of the algorithm.

Step 1. In the case 1.1 thesis of the theorem follows from lemma 2.17. In the case 1.2 there are two possibilities: either $|y| \leq |P|$ (this is step 1.2.1) or $|y| > |P|$ (this is step 1.2.2).

Step 2. It follows from the fact that there are three possibilities: either $|x| = |y|$ or $|x| < |y|$ or $|x| > |y|$.

Step 3. Steps 3.1 and 3.2 follow from proposition 3.1 of [8]. Step 3.3 follows from definition of a directed equation. The next two steps (3.4, 3.5) correspond to cases $|y| \leq |A|$ and $|y| > |A|$.

Step 4. By proposition 4.4 of [8] the equation $xAy\Phi = yBx\Psi$ is solvable if and only if there exists a word R of length c such that system of two equations $xAy = yBxR$, $R\Phi = \Psi$ is solvable.

Step 4.1 correspond to the case $|y| \leq c$. In this case we have $y = R_1$ for some suffix R_1 of R, and the first equation of our system gets the form $xAR_1 = R_1BxR$. Any solution of this equation must satisfy condition $x = (R_1B)^t R'$ for some $t \in \mathbb{N}$ and $R' < R_1B$.

Reading this equation from right to left we get $\overleftarrow{R}\,\overleftarrow{x}\,\overleftarrow{B}\,\overleftarrow{R_1} = \overleftarrow{R_1}\,\overleftarrow{A}\,\overleftarrow{x}$ where $\overleftarrow{P}$ is the reverse of the word P, i.e. the word P read from right to left. After reducing this equation we get $\overleftarrow{x}\,\overleftarrow{B}\,\overleftarrow{R_1} = \overleftarrow{A_i}\,\overleftarrow{x}$, where $i = |R_1|$. Any solution of this equation is of the form $\overleftarrow{x} = (\overleftarrow{A_i})^t \overleftarrow{A'}$ for some $t \in \mathbb{N}$ and $\overleftarrow{A'} < \overleftarrow{A_i}$ which means that $x = A'(A_i)^t$ where A' is some suffix of A_i. Since $|A_i| = |BR_1|$, we get $R_1BR' = A'A_i$, so $R_1 = R_{i,j}$ where $j = |A'|$.

Step 4.2 follows from proposition 5.10 of [8]. ■

Theorem 4.2 *The algorithm presented above works in polynomial time.*

Proof. This follows from the fact that in each step we create polynomial number of equations of polynomial length and from lemmas 2.3 and 2.8. ■

Remark. For the simplicity of the algorithm we did not take care of its complexity. In fact the complexity of the algorithm we have presented is of order $O(d^{100})$, what suggests that this algorithm is not of practical value. However, this complexity can be improved - namely in most places the number d can be replaced by a much smaller number called by Khmelevskiĭ the characteristics of an equation (see [8] for details).

5 An example

As an aplication of the algorithms given in sections 3 and 4, we shall solve the following problem.

Problem. Find all integers k, l such that the equation

$$(ax)^k = (yb)^l, \tag{8}$$

with coefficients $\Sigma = \{a, b\}$, and variables $\Xi = \{x, y\}$ is solvable. Describe the set of all solutions.

Without any loss of generality we can assume that $k \le l$, since otherwise we can consider the equivalent equation $(by)^l = (xa)^k$, obtained by reversing the order in (8). Therefore, it suffices to consider only solutions v, for which $|v(x)| \ge |v(y)|$. First, we shall follow the algorithm described in Section 4.

5.1 Reduction of the example to basic equations

Step 1 The equation has the form described by **Case 1.2**, so we follow instructions described in Step 1.2.1 and Step 1.2.2.
Step 1.2.1 We substitute a for y and get the equation

$$(ax)^k = (ab)^l$$

According to lemma 2.3 any solutions of this equation is of the form $v(x) = (ba)^t$ or $v(x) = (ba)^t b$. There is no solution of the first form since after substitution the left hand side of the equation ends with the symbol a while the right hand side ends with b. The second form provides a solution if $k(t+1) = l$, so we get the first solution of (8).

Lemma 5.1 *If $k|l$, then $v(y) = a$, $v(x) = b(ab)^{\frac{l}{k}-1}$ is a solution of $(ax)^k = (yb)^l$.*

Step 1.2.2 We substitute ay for y and get the equation

$$x(ax)^{k-1} = yb(ayb)^{l-1}$$

as an input for Step 2.
Step 2.1 We substitute y for x and get the

$$(ay)^{k-1} = b(ayb)^{l-1}$$

which is trivially unsolvable.

Since we are looking for solutions such that $|v(x)| \ge |v(y)|$, we do not follow Step 2.3. To execute Step 2.2 we go to Step 3.
Step 3 We are going to solve the directed equation $x(ax)^{k-1} \to yb(ayb)^{l-1}$.
Step 3.1 We substitute $(yb(ayb)^{l-1})^t Y$ for x where $Y \in \{\varepsilon, yb(ayb)^s, yb(ayb)^s a : s < l-1\}$. In this way we obtain the equation

$$(yb(ayb)^{l-1})^t Y [a(yb(ayb)^{l-1})^t Y]^{k-1} = yb(ayb)^{l-1}$$

By comparing lengths of both sides of the equation it is easy to check that if the equation is solvable, then $t = 0$. So, it remains to solve the equation $Y(aY)^{k-1} = yb(ayb)^{l-1}$ We consider three cases for different types of Y.

- $Y = \varepsilon$

  We get the equation
  $$a^{k-1} = yb(ayb)^{l-1}$$
  whose reverse is trivially unsolvable

- $Y = yb(ayb)^s$

  We get the equation
  $$yb(ayb)^s[ayb(ayb)^s]^{k-1} = yb(ayb)^{l-1},$$
  equivalent to $[(ayb)^{s+1}]^{k-1} = (ayb)^{l-s-1}$, which is solvable if and only if $(s+1)k = l$, and we get a second solution to (8).

Lemma 5.2 *If $k|l$, then $v(y) = ay$, $v(x) = yb(ayb)^{\frac{l}{k}-1}$ is a solution of $(ax)^k = (yb)^l$.*

- $Y = yb(ayb)^s a$

  We get the equation
  $$yb(ayb)^s a[ayb(ayb)^s a]^{k-1} = yb(ayb)^{l-1},$$
  which is equivalent to $a[(ayb)^{s+1}a]^{k-1} = (ayb)^{l-s-1}$. The reverse of the last equation is trivially unsolvable

Step 3.2 We substitute $(yb(ayb)^{l-1})^t Y$ for x where $Y \in \{x, yb(ayb)^s ax : s < l-1\}$. Again we get the equation
$$(yb(ayb)^{l-1})^t Y[a(yb(ayb)^{l-1})^t Y]^{k-1} \leftarrow yb(ayb)^{l-1},$$
and again we conclude that $t = 0$. Therefore the last equation reduces to $Y(aY)^{k-1} \leftarrow yb(ayb)^{l-1}$. Since $t = 0$, the case $Y = x$ gives the degenerated substitution of x for x, and it suffices to consider the case $Y = yb(ayb)^s ax$. Now, our equation has the form
$$yb(ayb)^s ax[ayb(ayb)^s ax]^{k-1} \leftarrow yb(ayb)^{l-1}$$
equivalent to $ax[(ayb)^{s+1}ax]^{k-1} = (ayb)^{l-s-1}$.
Step 3.3 We substitute xy for y and get $[(axyb)^{s+1}ax]^{k-1} = yb(axyb)^{l-s-2}$
Step 3.4 We substitute a for y and get
$$[(axab)^{s+1}ax]^{k-1} = ab(axab)^{l-s-2}.$$
This is equation of the form $x\Phi = bax\Psi$. We have to consider two cases

- Substitution of $(ba)^t$ for x gives a trivially unsolvable equation

- Substitution of $(ba)^t b$ for x gives the equation
  $$(ab)^{[(t+2)(s+1)+t+1](k-1)} = (ab)^{1+(t+2)(l-s-2)},$$
  which is solvable if and only if $k[(t+2)(s+1)+t+1] = l(t+2)$. Composing all the substitutions we have used, we get the third answer to our problem:

Lemma 5.3 *If there exist integers t, s, such that $k[(t+2)(s+1)+t+1] = l(t+2)$ and $s < l$ then, $v(y) = (ab)^{t+1}a$, $v(x) = b(ab)^{(t+2)(s+1)+t}$ is solution of $(ax)^k = (yb)^l$.*

Step 3.5 We substitute ay for y and get $[(axayb)^{s+1}ax]^{k-1} = ayb(axayb)^{l-s-2}$ as input of Step 4

Step 4.1 We have to solve equation of the form $xay\Phi = ybax\Psi$, so $c = 1$, $A_1 = ba$, $R_{1,0} = b$, $R_{1,1} = a$. Again, we consider two cases:

- Substitution of a for x gives an equation whose reverse is trivially unsolvable
- Substitution of b for x gives

  $$[(abayb)^{s+1}ab]^{k-1} = ayb(abayb)^{l-s-2}$$

  which is of the form $bay\Phi = y\Psi$. There are two possibilities of choosing solution:

  - Substitution of $(ba)^t b$ for y gives a trivially unsolvable equation
  - Substitution of $(ba)^t$ for y gives

    $$(ab)^{[(t+2)(s+1)+1](k-1)} = (ab)^{t+1+(t+2)(l-s-2)}$$

    which is solvable if and only if $k[(t+2)(s+1)+1] = l(t+2)$. Composing all the substitutions we have used, we get the fourth answer to our problem:

Lemma 5.4 *If there exist numbers t, s, such that $k[(t+2)(s+1)+1] = l(t+2)$ and $s < l$, then $v(y) = (ab)^{t+1}a$, $v(x) = b(ab)^{(t+2)(s+1)}$ is a solution of $(ax)^k = (yb)^l$.*

Step 4.2 The only possible decomposition is $P = S = \varepsilon$, $Q = A = ba$, so $R = b$, and we substitute xb for x. We get the equation

$$[(axbayb)^{s+1}axb]^{k-1} = ayb(axbayb)^{l-s-2}$$

of the form $xbay\Phi = ybax\Psi$, so we follow Case 1 of Algorithm 3.1.

5.2 Applying the algorithm for basic equations

Step 1.1 The only possible choice of Z, Z_1, Z_2 is $Z = ba$, $Z_1 = Z_2 = \varepsilon$. We substitute $(ba)^n$ for x and $(ba)^m$ for y. This substitution gives the equation

$$[(a(ba)^n ba(ba)^m b)^{s+1} a(ba)^n b]^{k-1} = a(ba)^m b[a(ba)^n ba(ba)^m b]^{l-s-2}$$

equivalent to $[(ab)^{(n+m+2)(s+1)+n+1}]^{k-1} = (ab)^{m+1+(n+m+2)(l-s-2)}$ which is solvable if and only if $k((n+m+2)(s+1)+n+1) = l(n+m+2)$. Here we get the fifth answer to our problem:

Lemma 5.5 *If there exist integers n, m, s such that $k((n+m+2)(s+1)+n+1) = l(n+m+2)$ and $s < l$, $n, m > 0$, then $v(y) = (ab)^{n+m+1}a$, $v(x) = b(ab)^{(n+m+2)(s+1)+n}$ is a solution of $(ax)^k = (yb)^l$.*

Steps 1.2 and **1.3** We substitute $(zba)^n z$ for x and $(zba)^m z$ for y thus obtaining the equation

$$[(azb)^{(n+m+2)(s+1)+n+1}]^{k-1} = (azb)^{m+1+(n+m+2)(l-s-2)},$$

which is solvable if and only if $k((n+m+2)(s+1)+n+1) = l(n+m+2)$. The last answer to our problem is:

Lemma 5.6 *If there exist nonnegative integers n, m, s, such that $k((n+m+2)(s+1)+n+1) = l(n+m+2)$ and $s < l'$ then $v(y) = (azb)^{n+m+1}az$, $v(x) = zb(azb)^{(n+m+2)(s+1)+n}$ is solution of equation $(ax)^k = (yb)^l$.*

Now we can recapitulate our solution:

Theorem 5.7 *Equation $(ax)^k = (yb)^l$ is solvable if and only if one of the four conditions below is satisfied:*

- *$k|l$ (solution: $y = az, x = zb(azb)^{\frac{l}{k}-1}$ for any $z \in \Sigma^*$)*
- *$l|k$ (solution: $x = az, y = zb(azb)^{\frac{k}{l}-1}$ for any $z \in \Sigma^*$)*
- *there exist nonnegative numbers n, m, s such that $k((n+m+2)(s+1)+n+1) = l(n+m+2)$ and $s < l$ (solution: $y = (azb)^{n+m+1}az, x = zb(azb)^{(n+m+2)(s+1)+n}$ for any $z \in \Sigma^*$)*
- *there exist nonnegative numbers n, m, s such that $l((n+m+2)(s+1)+n+1) = k(n+m+2)$ and $s < k$ (solution: $x = (azb)^{n+m+1}az, y = zb(azb)^{(n+m+2)(s+1)+n}$ for any $z \in \Sigma^*$)*

References

[1] H. Abdulrab, Résolution d'équations sur les mots: étude et implémentation LISP de l'algorithme de Makanin. *Ph.D. Thesis, Université de Rouen*, 1987.

[2] H. Abdulrab, J.P. Pecuchet, Solving word equations. *Journal of Symbolic Computation* 8 (1989), pp. 499-521.

[3] S.I. Adyan, G.S. Makanin, Investigation on algorithmic questions of algebra. *Trudy Matem. Inst. Steklova* 168 (1984), English translation in *Proc. of Steklov Institute of Mathematics* 1986, issue 3, pp. 207-226.

[4] V.K. Bulitko, Equations and inequalities in a free group and a free semigroup. (in Russian) *Tul. Gos. Ped. Inst. Ucen. Zap. Mat. Kafedr Vyp 2, Geometr. i Algebra.* (1970), pp.242-252.

[5] W. Charatonik, Equations in Free Semigroup (in Polish),*Master Thesis, Wroclaw University* 1991 (unpublished).

[6] J. Jaffar, Minimal and Complete Word Unification. *Journal of ACM* 37 (1990), pp.47-85.

[7] Yu.I. Khmelevskiĭ, Solution of word equations in three unknowns. *Dokl. Akad. Nauk SSSR* 177 (1967), 1023-1025; English transl. in *Soviet Math. Dokl.* 8 (1967).

[8] Yu.I. Khmelevskiĭ, Equations in a Free Semigroup (in Russian), *Trudy Matem. Inst. Steklova*, 107 (1971) pp. 1–284.

[9] A. Kościelski, L. Pacholski, Complexity of Unification in Free Groups and Free Semigroups, *Proceedings 31st Annual Symposium on Foundations of Computer Science*, Los Alamitos 1990, vol. II, pp.824–830.

[10] A. Kościelski, L. Pacholski, Complexity of Makanins' Algorithms *submitted.*

[11] A. Lentin, Equations in free monoids. in Nivat, editor, *Automata, Languages and Programing,* Amsterdam 1972, pp. 67-85.

[12] G.S. Makanin, The Problem of Solvability of Equations in a Free Semigroup. (in Russian), *Matematicheskiĭ Sbornik* 103 (1977), pp. 147-236; English translation in *Math. USSR Sbornik* 32 (1977), pp. 129-198.

[13] J.-P. Pecuchet, Solutions principales et rang d'un système d'équations avec constantes dans le monoïde libre. *Discrete Mathematics* 48 (1984), pp.253-274.

[14] G. Plotkin, Building in equational theories. *Machine Intelligence* 7 (1972), pp. 73-90.

[15] K.U. Schulz, Makanin's algorithm - two improvements and a generalization, *CIS-report* 91-39, Centrum für Informations- und Sprachverarbeitung, University of Munique, 1991.

The Naming Problem for Left Distributivity.

PATRICK DEHORNOY

ABSTRACT. We consider the problem of naming the variables of a term so that it becomes equivalent to a given term when left distributivity is assumed, and describe an algorithm for solving this question using conjugacy in a free group. The correctness of the algorithm is reduced to a conjecture involving some particular words. A skew version of the conjecture is established.

1. The Naming Problem for left distibutivity.

Assume that Σ is a fixed set whose elements will be called variables. We denote by $\mathcal{T}_\Sigma$ the set of all wellformed terms constructed from Σ using a single binary operation. It will be convenient to use a bracket notation for this operation, thus writing $P[Q]$ for the product of the terms P and Q.

Let $\approx$ denote the least congruence on $\mathcal{T}_\Sigma$ for which

$$P[Q[R]] \approx P[Q][P[R]]$$

holds for all terms P, Q, R. Thus $\mathcal{T}_\Sigma/\approx$ is exactly the free left distributive magma generated by Σ. In this paper we are interested in the following problem. Assume that we are given two terms P, Q. We ask whether by changing the names of the variables in Q we can make it $\approx$-equivalent to P. Formally we may assume that the variables in Q are pairwise distinct (we shall say that Q is an *injective* term), and the question is whether there exists a mapping σ of the variables occurring in Q into the variables occurring in P such that Q^σ is $\approx$-equivalent to P, where Q^σ is the term obtained from Q by replacing every variable x occurring in Q by the corresponding variable x^σ (*i.e.* $\sigma(x)$). This question will be called the *Naming Problem* for the pair (P, Q) in the sequel.

Example. Assume that P is $a[b[c[d]]]$ and Q is $A[B[C][D][E[F][G]]]$. Then the substitution σ defined by

$$A^\sigma = a, \quad B^\sigma = b, \quad C^\sigma = c, \quad D^\sigma = b, \quad E^\sigma = b, \quad F^\sigma = c, \quad G^\sigma = d$$

is a solution to the Naming Problem for (P, Q), since the term Q^σ is $a[b[c][b][b[c][d]]]$, which is $\approx$-equivalent to P.

It will be useful to consider terms as binary trees whose leaves are labelled using variables in Σ. We use finite sequences of 0's and 1's as addresses for the nodes in binary trees (starting from the root, whose address is the empty sequence Λ). The set of all addresses, *i.e.* the free monoid generated by 0 and 1, will be denoted by $\mathbf{S}$. For P in $\mathcal{T}_\Sigma$, the set of all addresses of leaves in the associated tree is called the *support* of the term P, and denoted by Supp$P$. For instance the support of the term $a[b[c[d]]]$ is the set $\{0, 10, 110, 111\}$. For w in the support of P, we simply write $P(w)$ for the variable x which occurs at w in P. In this case we say that w is an occurrence of x in P.

The problem above can be stated in geometrical terms as follows. Assume that we are given a term (*i.e.* a labelled tree) and a tree. The question is to find a variable assignment for the tree so that the term thus obtained is $\approx$-equivalent to the original term. For instance the problem in the example above can be illustrated by the following picture

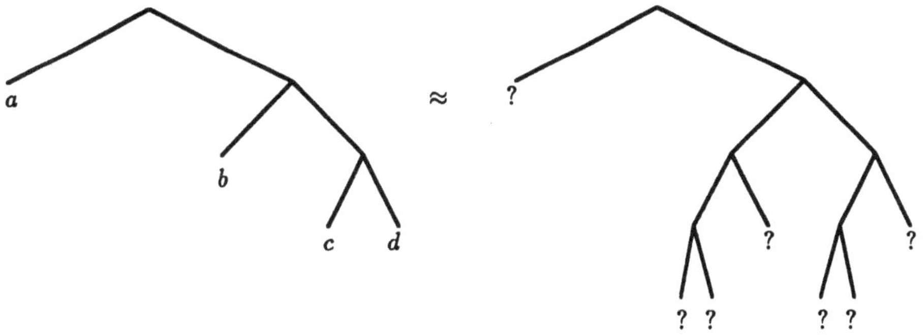

The properties of left distributivity imply the solvability of the Naming Problem. First we have

Proposition 1.- *For a given pair of terms (P, Q), the Naming Problem has at most one solution.*

Proof. Let us say that two terms are isometric when they have the same support. The point is that two isometric terms Q', Q'' can only be either equal or $\approx$-unequivalent. For if Q', Q'' are not equal, consider the leftmost discrepancy between them. Using the technics of [2], one finds new terms R', R'' which are extensions of Q' and Q'' respectively and are such that some iterated left subterms S', S'' of R', R'' respectively coincide except for their righmost variables. One deduces that either S' is equivalent to a left subterm of S'', or that S'' is equivalent to a left subterm of S'. In both cases one obtains a contradiction to the socalled irreflexivity property, which claims that a term cannot be equivalent to one of its (iterated) subterms. This property was first established by Richard Laver in [7] using an auxiliary set theoretical hypothesis, and subsequently in [5] without any logical assumption. □

The solution to the Naming Problem need not exist in general. For instance if P is a single variable, then $P \approx Q^\sigma$ has no solution when the support of Q has more than one point, since the only term $\approx$-equivalent to P is P itself.

Proposition 2.- *For every pair of terms (P, Q), the Naming Problem for (P, Q) is decidable in primitive recursive time.*

Proof. The Word Problem for $\approx$ on $\mathcal{T}_\Sigma$ is known to be decidable with a primitive recursive complexity (*c.f.* [5]). We can then solve the Naming Problem for (P, Q) by systematically enumerating all possible substitutions σ of the variables of Q into the variables of P and checking for each of them whether Q^σ is $\approx$-equivalent to P. □

We would like to obtain a better algorithm. In [3] a conjectural algorithm is described for the Word Problem of $\approx$, and this algorithm can be easily modified to solve the Naming Problem as well. Unfortunately this algorithm, which is known to be correct since [5], is not yet known to be convergent.

Before introducing a new algorithm, let us consider a more particular problem. We say that a term Q is an *extension* of the term P, and write $P \rightarrow Q$, if Q is obtained from P by iteratively replacing some subterm of the form $P'[P''[P''']]$ by the corresponding term $P'[P''][P'[P''']]$. It is clear that $P \rightarrow Q$ implies $P \approx Q$, and $\approx$ is easily shown to be the equivalence relation generated by $\rightarrow$. It was proved in [2] that the relation $\rightarrow$ is confluent, which implies that $P \approx Q$ holds exactly if and only if for some term R both $P \rightarrow R$ and $Q \rightarrow R$ hold (the Church-Rosser property).

Assume again that (P, Q) is a pair of terms and that Q is injective. The *Restricted* Naming Problem for (P, Q) will be the question of whether there exists a substitution σ such that the term Q^σ is an extension of the term P (which implies that Q^σ is $\approx$-equivalent to P).

The Restricted Naming Problem is easier than the general one. To see that introduce the following notations. If P is a term in $\mathcal{T}_\Sigma$, and if w is a point in $\mathbf{S}$, we write $P_{/w}$ for the subterm of P with root in w if it exists, *i.e.* if w is short enough. We write Ω_w for the partial operator on $\mathcal{T}_\Sigma$ which maps a term P to the term Q obtained by replacing the subterm $P_{/w}$, *i.e.* $P_{/w0}[P_{/w10}[P_{/w11}]]$, by $P_{/w0}[P_{/w10}][P_{/w0}[P_{/w11}]]$ (when possible, *i.e.* when all these subterms exist). Thus Ω_w performs the elementary distribution transformation at w. Clearly a term Q is an extension of the term P if there exists a finite sequence $(w_1, \ldots, w_n)$ in $\mathbf{S}^*$ such that the composition of Ω_{w_1}, $\ldots$, Ω_{w_n} maps P to Q.

Proposition 3.- *For every pair of terms* (P, Q), *the Restricted Naming Problem for* (P, Q) *is decidable in polynomial space.*

Proof. Since every transformation Ω_w strictly increases the size of the terms, we know that, if Q^σ is an extension of P, then Q^σ is obtained from P by applying a sequence of elementary transformations Ω_w whose length is at most the difference between the sizes of Q and P. Fix any linear ordering on $\mathbf{S}$, and use its lexicographical extension to $\mathbf{S}^*$ to enumerate all possible sequences of elementary transformations which may be applied to P. When the size goes beyond the size of Q, stop exploring this branch of $\mathbf{S}^*$ and go to the next one. One has only to record the current extension of P (a term whose size is at most the size of Q), and the current sequence of addresses. The length of this sequence is, by the remark above, at most the size of Q, and each element of the sequence is an address which lies "inside of Q" and its length (as a sequence of 0's and 1's) is certainly bounded by the size of Q too. So the needed space lies in $O(\text{size}(Q)^2)$. □

Example. Assume that P is $a[b[c[d]]]$ and Q is $A[B[C][D][E[F][G]]]$. Use the natural ordering on $\mathbf{S}$ which satisfies $\Lambda < 0 < 1$. The successive steps of the preceding algorithm entail the following sequences of addresses and extensions of P.

step 1	(Λ)	$a[b][a[c[d]]]$	continue
step 2	(Λ, Λ)	$a[b][a][a[b][c[d]]]$	continue
step 3	$(\Lambda, \Lambda, \Lambda)$	$a[b][a][a[b]][a[b][a][c[d]]]$	backtrack
step 4	$(\Lambda, \Lambda, 1)$	$a[b][a][a[b][c][a[b][d]]]$	backtrack
step 5	$(\Lambda, 1)$	$a[b][a[c][a[d]]]$	continue
step 6	$(\Lambda, 1, \Lambda)$	$a[b][a[c]][a[b][a[d]]]$	backtrack
step 7	$(\Lambda, 1, 1)$	$a[b][a[c][a][a[c][d]]]$	backtrack
step 8	(1)	$a[b[c][b[d]]]$	continue
step 9	$(1, \Lambda)$	$a[b[c]][a[b[d]]]$	continue
step 10	$(1, \Lambda, \Lambda)$	$a[b[c]][a][a[b[c]][b[d]]]$	backtrack
step 11	$(1, \Lambda, 0)$	$a[b][a[c]][a[b[d]]]$	backtrack
step 12	$(1, \Lambda, 1)$	$a[b[c]][a[b][a[d]]]$	backtrack
step 13	$(1, 1)$	$a[b[c][b][b[c][d]]]$	success

and we finally obtain the desired solution.

2. The Conjugacy Bracket.

The aim of this paper is to present a better algorithm for solving the Naming Problem by using the conjugacy in free groups. Assume that $(G,.)$ is any group. The conjugacy bracket defined on G by

$$x[y] = xyx^{-1}$$

is left distributive. The idea is to use the values of the terms in a group G using the conjugacy operation to compute the bracket. Because conjugacy is left distributive, any substitution σ such that Q^σ and P are $\approx$-equivalent will yield equal evaluations for the terms Q^σ and P in the group G. So we shall try to find σ by directly using the values in G.

To avoid losing information as much as possible when projecting, we shall use the most general possible group, namely the free group. In the sequel we denote by $\mathcal{G}_\Sigma$ the free group generated by Σ. The elements of $\mathcal{G}_\Sigma$ are represented by words constructed from the elements of Σ and their inverses. The set of all such words, *i.e.* the free monoid generated by Σ and of a disjoint copy $\overline{\Sigma}$ of Σ, is denoted by $\mathcal{W}_\Sigma$. Thus $\mathcal{G}_\Sigma$ is the quotient of the monoid $\mathcal{W}_\Sigma$ under the congruence $\equiv$ generated by the pairs $\{x\overline{x}, \varepsilon\}$ and $\{\overline{x}x, \varepsilon\}$, where ε is the empty word and $\overline{\alpha}$ is the word obtained from α by reversing the ordering of the characters and permuting x and $\overline{x}$ for every x in S. It is wellknown that every class for $\equiv$ contains exacly one reduced word, *i.e.* a word where no pattern $x\overline{x}$ or $\overline{x}x$ appears. For α in $\mathcal{W}_\Sigma$, the unique reduced word which is $\equiv$-equivalent to α will be denoted by $\mathrm{red}(\alpha)$.

Words will be considered as finite sequences, *i.e.* mappings whose domain is an integer interval of the form $1..n$. If α is a word of length n, the interval $1..n$ will simply be denoted by $\mathrm{Dom}(\alpha)$, and for $1 \leq k \leq n$ the k-th character in α will be denoted by $\alpha(k)$. In this case we shall say that k is an occurrence of the character $\alpha(k)$ in α. If α is a word in $\mathcal{W}_\Sigma$ and k belongs to $\mathrm{Dom}(\alpha)$, we say that k is a ***positive*** occurrence in α if $\alpha(k)$ belongs to Σ, and a ***negative*** occurrence if $\alpha(k)$ belongs to $\overline{\Sigma}$.

Definition.- For P in $\mathcal{T}_\Sigma$, $\langle P\rangle$ is the word in $\mathcal{W}_\Sigma$ inductively defined by

$$\langle P\rangle = \begin{cases} P & \text{if } P \text{ lies in } \Sigma, \\ \langle P_0\rangle\langle P_1\rangle\overline{\langle P_0\rangle} & \text{if } P \text{ is } P_0[P_1]. \end{cases}$$

Thus (if $\mathcal{G}_\Sigma$ is identified with the set of all reduced words) the set $\mathcal{C}_\Sigma$ of all "pure conjugates" $\mathrm{red}\langle P\rangle$ for P in $\mathcal{T}_\Sigma$ is the closure of Σ in $\mathcal{G}_\Sigma$ using the conjugacy bracket. Clearly $\mathcal{C}_\Sigma$ is a strict subset of $\mathcal{G}_\Sigma$, and is not a subgroup of $\mathcal{G}_\Sigma$, since if a, b are distinct elements of Σ, the word ab is reduced and cannot belong to $\mathcal{C}_\Sigma$ (see Lemma 2 below for a precise argument).

Example. If P is the term $a[b][c[d]]$, the word $\langle P\rangle$ is the word $a\,b\,\bar{a}\,c\,d\,\bar{c}\,a\,\bar{b}\,\bar{a}$.

We summarize now a few properties of the words $\langle P\rangle$ and $\mathrm{red}\langle P\rangle$.

Lemma 1.- *i) If P, Q are $\approx$-equivalent terms, the words $\langle P\rangle$ and $\langle Q\rangle$ are $\equiv$-equivalent.*

ii) If P is an injective term, the word $\langle P\rangle$ is a reduced word.

Proof. Induction on (the size of) the term P. If P is a variable the result is obvious. If P is $P_0[P_1]$, the variables in P_0 and P_1 are assumed to be distinct, so that no cancellation may happen between the words $\langle P_0\rangle$ and $\langle P_1\rangle$, or between $\langle P_1\rangle$ and $\overline{\langle P_0\rangle}$. □

Of course this result is not true when P is not injective, and the converse of Lemma 1 is false. For example a and $a[a]$ are $\approx$-unequivalent terms, but $\langle a\rangle$ and $\langle a[a]\rangle$ are equal to a.

Definition.- A word α in $\mathcal{W}_\Sigma$ with length n is *antisymmetric* if n is an odd number, if for every k between 1 and $(n-1)/2$ one has

$$\alpha(n-k) = \overline{\alpha(k)},$$

and moreover if the occurrence $(n+1)/2$ is positive in α.

Lemma 2.- *For any term P the words $\langle P\rangle$ and $\mathrm{red}\langle P\rangle$ are antisymmetric.*

Proof. If α, β are antisymmetric words, so is the word $\alpha\beta\bar{\alpha}$. Because $\langle P\rangle$ is antisymmetric when P is a single variable, this inductively implies that $\langle P\rangle$ is antisymmetric for every term P. Now we claim that the reduction of an antisymmetric word yields an antisymmetric word. For assume that α is antisymmetric. Let $2n+1$ be the length of α. Assume that α' is obtained from α using one step of elementary reduction. Thus α' is obtained by deleting two consecutive occurrences say k and $k+1$ in α. Assume that k is a positive occurrence and $k+1$ is a negative one. So for some x in Σ, one has $\alpha(k) = x$ and $\alpha(k+1) = \bar{x}$. First assume that k is not $n+1$. Then $k+1 = n+1$ is impossible by definition, and the pair $\{2n+1-k, 2n-k\}$ is disjoint from the pair $\{k, k+1\}$. By definition one has $\alpha(2n+1-k) = \bar{x}$ and $\alpha(2n-k) = x$. Hence α' can be subsequently reduced by cancelling the pair in α' which comes from $2n+1-k$ and $2n-k$, and the word α'' thus obtained is certainly antisymmetric. Now assume that k is $n+1$. Because $\alpha(k+1)$ is $\bar{x}$, $\alpha(k-1)$ must be x, and it follows that the word α' is still antisymmetric (the central occurrence is now n, which is positive). The argument is similar if k is negative in α. □

For improving the analysis, we introduce for every occurrence in a word an integer level as follows.

Definition.- For α in $\mathcal{W}_\Sigma$, the mapping $\alpha\hat{}: \mathrm{Dom}(\alpha) \to \mathbf{Z}$ is defined by

$$\alpha\hat{}(k) = \begin{cases} 0 & \text{if } k = 1, \\ \alpha(k-1)\hat{} + 1 & \text{if } k > 1,\ \alpha(k-1) \in \Sigma \text{ and } \alpha(k) \in \Sigma, \\ \alpha\hat{}(k-1) - 1 & \text{if } k > 1,\ \alpha(k-1) \in \overline{\Sigma} \text{ and } \alpha(k) \in \overline{\Sigma}, \\ \alpha\hat{}(k-1) & \text{otherwise.} \end{cases}$$

Example.- Let Q be the term $A[B[C][D][E[F][G]]]$ as above. Then the word $\langle Q \rangle$ when written "in two dimensions" for emphasizing the level of each occurrence becomes

level 4 →													E	$\overline{F}$			C	$\overline{B}$					
level 3→						B	$\overline{C}$			F	$\overline{E}$	G			$\overline{E}$	B			$\overline{D}$	B	$\overline{C}$		
level 2 →			C	$\overline{B}$	D			$\overline{B}$	E													$\overline{B}$	
level 1→		B																					$\overline{A}$
level 0 →	A																						

Lemma 3.- *i) In any word $\langle P \rangle$ the level of every occurrence except the first one is at least 1. If P is not a single variable the last occurrence in $\langle P \rangle$ is negative at level 1.*

ii) For any terms P, Q, the following formula

$$\langle P[Q] \rangle\hat{}(k) = \begin{cases} \langle P \rangle\hat{}(k) & \text{if } 1 \le k \le \lg\langle P \rangle, \\ \langle Q \rangle\hat{}(\ell) + 1 & \text{if } k = \lg\langle P \rangle + \ell \text{ with } 1 \le \ell \le \lg\langle Q \rangle, \\ \langle P \rangle\hat{}(\lg\langle P \rangle - \ell) + 1 & \text{if } k = \lg\langle P \rangle + \lg\langle Q \rangle + \ell \text{ with } 1 \le \ell \le \lg\langle P \rangle \end{cases}$$

holds.

Proof. The first occurrence in any word $\langle P \rangle$ is positive at level 0. So if the property for the last occurence holds for P and Q, the definition of $\langle P[Q] \rangle$ gives the formula for $\langle P[Q] \rangle\hat{}$, which in turn implies the property of (i) for the term $P[Q]$. So the proof goes on inductively. □

Proposition 4.- *i) The level of every occurrence except 1 in the word* $\mathrm{red}\langle P \rangle$ *is at least 1. If* $\mathrm{red}\langle P \rangle$ *has length p, then the relation*

$$\mathrm{red}\langle P \rangle\hat{}(p - k) = \mathrm{red}\langle P \rangle\hat{}(k) + 1$$

holds for $1 \le k < p/2$.

ii) The first occurrence in $\langle P \rangle$ cannot disappear when this word is reduced. The central occurrence in $\mathrm{red}\langle P \rangle$ *is an occurrence of the rightmost variable in P.*

Proof. It suffices to observe that word reduction preserves the level of the remaining occurrences, and that an occurrence in a word can possibly be cancelled only with an occurrence at the same level. This is true since by very definition two consecutive occurrences with opposite signs must have the same level, and that deleting a positive-negative or a negative-positive pair of occurrences does not change the levels of the other occurrences. □

Remark.- The preservation of level under reduction implies that the levels of occurrences in words red$\langle P\rangle$ are preserved when P is replaced by an extension of P. This can be explained as follows. Code the operation of increasing (*resp.* decreasing) level by $\uparrow$ (*resp.* $\overline{\uparrow}$). For instance the code of the two-dimensional writing of the word $\langle Q\rangle$ for Q as above will be the word $\langle Q\rangle'$ beginning with $A\uparrow B\uparrow C\overline{B}D\uparrow B\overline{C}\,\overline{\uparrow}\overline{B}\ldots$. Then $\langle Q\rangle'$ admits the same inductive definition as $\langle Q\rangle$ but with a bracket now defined by

$$x[y] = x\uparrow y\overline{x}\,\overline{\uparrow}.$$

The point is that this new bracket is still left distributive, as is any bracket defined by $x[y] = xf(y\overline{x})$ where f is a group endomorphism.

When going from a term P to the word $\langle P\rangle$, every occurrence in P, *i.e.* every point in the set Supp$P$, gives rise to several occurrences in $\langle P\rangle$ in general. For w in Supp$P$, we denote by Im(P,w) the set of all such occurrences. Formally $w\mapsto$ Im(P,w) is the mapping of Supp$P$ into $\mathfrak{P}(\mathrm{Dom}\langle P\rangle)$ defined by $\mathrm{Im}(x,\Lambda)=1$ for x in Σ and by

$$\mathrm{Im}(P,w) = \begin{cases} \mathrm{Im}(P_0,w_0)\cup(\mathrm{Im}(P_0,w_0)+\mathrm{lg}\langle P_0\rangle+\mathrm{lg}\langle P_1\rangle) & \text{if } w=0w_0,\\ \mathrm{Im}(P_1,w_1)+\mathrm{lg}\langle P_0\rangle & \text{if } w=1w_1,\end{cases}$$

where P is $P_0[P_1]$ and $A+n$ means $\{k+n; k\in A\}$.

Lemma 5.- *Assume that w belongs to* Supp$P$, *and that w contains p times 0 and q times 1.*

i) The set Im(P,w) *has 2^p elements and $k\in \mathrm{Im}(P,w)$ is equivalent to $n-k\in \mathrm{Im}(P,w)$ where n is the length of $\langle P\rangle$.*

ii) The elements of Im(P,w) *are alternatively occurrences of $P(w)$ and $\overline{P(w)}$ (beginning with $P(w)$).*

iii) The level of the elements of Im(P,w) *in $\langle P\rangle$ are $q+\nu_1$, $q+\nu_2$, etc... where $(\nu_i)_{i\geq 1}$ is the "folding" sequence inductively defined by*

$$\nu_{2^k+i} = \nu_{2^k-i+1}+1 \textit{ for } 1\leq i\leq 2^k$$

from $\nu_1=0$.

Example.- If Q is the term $A[B[C][D][E[F][G]]]$, then 1000 is an occurrence of B in Q. The set Im$(Q,1000)$ has 8 elements, namely 2, 4, 6, 8, 16, 18, 20, 22. Their respective levels are 1, 2, 3, 2, 3, 4, 3, 2.

Proposition 6.- *The mapping $P\mapsto\langle P\rangle$ is injective.*

Proof. It suffices to show how the words $\langle P\rangle$ and $\langle Q\rangle$ can be obtained from the word $\langle P[Q]\rangle$. Now by Lemma 3 the first occurrence in $\langle P[Q]\rangle$ which comes from $\langle Q\rangle$ has level 1, while any further occurrence before the middle of $\langle P[Q]\rangle$ has level at least 2. □

To conclude these remarks we may observe that the numbers of occurrences of a variable on each level are easily characterized. In analogy with the free differential calculus (*c.f.* [1] for instance) we make the following

Definition.- The *Alexander polynomial* of a word α is the polynomial $\Delta_\alpha(X)$ with coefficients in $\mathbb{Z}[\Sigma]$ defined by

$$\Delta_\alpha(X) = \sum_{h=0}^{\infty}(\sum_{x\in\Sigma} \mu_{x,h}(\alpha)x)X^h$$

where

$$\mu_{x,h}(\alpha) = \text{card}\{k; \alpha(k) = x \text{ and } \alpha\hat{}(k) = h\} - \text{card}\{k; \alpha(k) = \overline{x} \text{ and } \alpha\hat{}(k) = h\}.$$

By the proof of Proposition 5 observe that the Alexander polynomials of two $\equiv$-equivalent words must coincide. In particular the polynomials $\Delta_\alpha(X)$ and $\Delta_{\text{red}\,\alpha}(X)$ always coincide.

Example.- The Alexander polynomial is obtained by counting on each level the difference between positive and negative occurrences of each variable. It is a kind of abelianization of the word. If Q is the term $A[B[C][D][E[F][G]]]$ as above, the polynomial $\Delta_{\langle Q\rangle}(X)$ is

$$\begin{aligned}A + (B - A)X + (E + D + C - 3B)X^2 + \\ (G + F - 2E - D - 2C + 3B)X^3 + (-F + E + C - B)X^4.\end{aligned}$$

Proposition 7.- *For every term P in T_Σ, the polynomial $\Delta_{\langle P\rangle}(X)$ is the value of the term P in the left distributive magma $\mathbb{Z}[\Sigma][X]$ equipped with the bracket*

$$P[Q] = (1 - X)P + XQ.$$

Proof. Induction on the term. For x in Σ, the polynomial $\Delta_{\langle x\rangle}(X)$ is the constant x. Now by Lemma 3 we have the formulas

$$\mu_{x,h}(\langle P[Q]\rangle) = \mu_{x,h}(\langle P\rangle) + \mu_{x,h-1}(\langle Q\rangle) - \mu_{x,h-1}(\langle P\rangle)$$

which give the induction relation

$$\Delta_{\langle P[Q]\rangle}(X) = (1 - X)\Delta_{\langle P\rangle}(X) + X\Delta_{\langle Q\rangle}(X).\square$$

3. The Group Algorithm.

Assume that P, Q are injective terms with disjoint variables. If the substitution σ is a solution for the associated Naming Problem, *i.e.* if Q^σ is $\approx$-equivalent to P, then $\langle P\rangle \equiv \langle Q^\sigma\rangle$ holds by Lemma 2.1. So the word $\overline{\langle P\rangle}\langle Q^\sigma\rangle$ reduces to the empty word. Hence a possible approach to the problem is to directly find a substitution σ such that $\overline{\langle P\rangle}\langle Q^\sigma\rangle$ is a fully reducible word. It is not true however that any solution to the latter question gives a solution for the Naming Problem. For instance if P is a and Q is $A[B]$, then the word $\overline{\langle P\rangle}\langle Q\rangle$ is $\bar{a}\,A\,B\,\bar{A}$, and the substitution σ defined by $A^\sigma = B^\sigma = a$ lets the word $\overline{\langle P\rangle}\langle Q^\sigma\rangle$ vanish, while the terms a and $a[a]$ are not $\approx$-equivalent.

Definition.- The word α is *solvable* with respect to the variable x if

i) x occurs in α,

ii) the first occurrence k of x or $\bar{x}$ in α is an occurrence of x,

iii) k is at least 2 and $k-1$ is a negative occurrence in α.

If α is solvable with respect to x, the (partial) substitution of Σ which maps x to the unique variable y such that the first occurrence of x is preceded by an occurrence of $\bar{y}$ (*i.e.* the pair (x, y)) is denoted by $\varphi(\alpha, x)$.

Example.- The word $a\,\bar{b}\,a\,A\,B\,\bar{A}$ is solvable with respect to A, and $\varphi(a\,\bar{b}\,a\,A\,B\,\bar{A}, A)$ is the mapping $A \mapsto a$. This word is not solvable with respect to B.

We are now ready to describe our algorithm.

Algorithm ($\mathcal{A}$).

Input. Two terms P, Q in $\mathcal{T}_\Sigma$ with disjoint variables.

Action. Set $\alpha_0 = \overline{\langle P\rangle}\langle Q\rangle$ and $\sigma_0 = \emptyset$. Let $y_1, \ldots, y_n$ be the variables in Q enumerated from left to right. For $i = 1$ to $i = n$ repeat the following step: if α_{i-1} is solvable with respect to y_i, then set

$$\begin{cases} \alpha_i = \mathrm{red}({\alpha_{i-1}}^{\varphi(\alpha_{i-1}, y_i)}) \\ \sigma_i = \sigma_{i-1} \cup \varphi(\alpha_{i-1}, y_i). \end{cases}$$

Output. The substitution σ_n, if no obstruction happened and the word α_n is empty.

Example.- Start again with $P = a[b[c[d]]]$ and $Q = A[B[C][D][E[F][G]]]$ as in Section 1. The successive steps of the algorithm ($\mathcal{A}$) performed from P and Q correspond to

$$\begin{array}{rl} & \alpha_0 = a\;b\;c\;\bar{d}\;\bar{c}\;\bar{b}\;\bar{a}\;ABC\bar{B}DB\bar{C}\bar{B}EF\bar{E}GE\bar{F}\bar{E}BC\bar{B}\bar{D}B\bar{C}\bar{B}\bar{A}, \\ A \mapsto a & \alpha_1 = a\;b\;c\;\bar{d}\;\bar{c}\;\bar{b}\quad BC\bar{B}DB\bar{C}\bar{B}EF\bar{E}GE\bar{F}\bar{E}BC\bar{B}\bar{D}B\bar{C}\bar{B}\,\bar{a}, \end{array}$$

$$\begin{array}{lll}
B \mapsto b & \alpha_2 = a\ b\ c\ \overline{d}\ \overline{c} & C\,\overline{b}\,D\,b\,\overline{C}\,\overline{b}\,E\,F\,\overline{E}\,G\,E\,\overline{F}\,\overline{E}\,b\,C\,\overline{b}\,\overline{D}\,b\,\overline{C}\,\overline{b}\,\overline{a}, \\
C \mapsto c & \alpha_3 = a\ b\ c\ \overline{d} & \overline{b}\,D\,b\,\overline{c}\,\overline{b}\,E\,F\,\overline{E}\,G\,E\,\overline{F}\,\overline{E}\,b\,c\,\overline{b}\,\overline{D}\,b\,\overline{c}\,\overline{b}\,\overline{a}, \\
D \mapsto b & \alpha_4 = a\ b\ c\ \overline{d} & b\,\overline{c}\,\overline{b}\,E\,F\,\overline{E}\,G\,E\,\overline{F}\,\overline{E}\,b\,c\,\overline{b}\quad \overline{c}\,\overline{b}\,\overline{a}, \\
E \mapsto b & \alpha_5 = a\ b\ c\ \overline{d} & b\,\overline{c}\quad F\,\overline{b}\,G\,b\,\overline{F}\quad c\,\overline{b}\quad \overline{c}\,\overline{b}\,\overline{a}, \\
F \mapsto c & \alpha_6 = a\ b\ c\ \overline{d} & G \quad \overline{c}\,\overline{b}\,\overline{a}, \\
G \mapsto d & \alpha_7 = \varepsilon. &
\end{array}$$

The algorithm succeeds and produces the substitution

$$\sigma : A \mapsto a, B \mapsto b, C \mapsto c, D \mapsto d, E \mapsto b, F \mapsto c, G \mapsto d.$$

This substitution is the solution to the Naming Problem for P and Q which has already been obtained above, since Q^σ is the extension $a[b[c][b][b[c][d]]]$ of P.

The method of the algorithm $(\mathcal{A})$ is quite natural: the substitution is constructed in such a way that each new value induces a new cancellation. The specific point is that we emphasize the left-right ordering of the variables and only consider the cancellations arising from the leftmost possible occurrences. For instance the algorithm $(\mathcal{A})$ will not work for the terms a and $A[B]$, since, starting from the word $\overline{a}\,A\,B\,\overline{A}$, one first infers $A \mapsto a$ thus obtaining the reduced word $B\,a$, which will be rejected although the further substitution $B \mapsto a$ could of course be applied to cancel it subsequently.

For u, v in $\mathbf{S}$, we write $u \sqsubset v$ if u is a prefix of v, *i.e.* il v is uw for some w. We write $u < v$ if u is on the left of v, *i.e.* if $w0 \sqsubset u$ and $w1 \sqsubset v$ hold for some w. If P is a term, the set of all prefixes of points in Supp$P$ is denoted by Supp^+P: a point in Supp^+P is the address either of a leaf of the tree associated with P or of an inner node of this tree. Now for w in Supp^+P there always exist an unique integer h such that $w1^h$ belongs to Supp$P$. We extend the former notations so that, in this case, $P(w)$ denotes $P(w1^h)$ and $\text{Im}(P, w)$ denotes $\text{Im}(P, w1^h)$. Thus $P(w)$ is always the rightmost variable in the subterm $P_{/w}$.

Definition.- For u, v in Supp^+P with $u < v$, $\langle P\rangle_u^v$ denotes the subword of $\langle P\rangle$ which begins with the first occurrence in $\text{Im}(P, u)$ and ends immediately before the first occurrence in $\text{Im}(P, v)$. If the index u is dropped, one begins with the first character of $\langle P\rangle$. If the exponent v is dropped, one goes until the end of $\langle P\rangle$.

Example.- Let Q be $A[B[C][D][E[F][G]]]$ again. Then $\langle Q\rangle_{1000}^{1100}$ is the word $B\,C\,\overline{B}\,D\,B\,\overline{C}\,\overline{B}$, while $\langle Q\rangle_{1100}^{\Lambda}$ is $E\,F\,\overline{E}$.

Observe that the formula $\langle P\rangle_u^w = \langle P\rangle_u^v \langle P\rangle_v^w$ always holds if $u < v < w$ is true (and u, v, w lie in Supp^+P), and that

$$\langle P\rangle = \langle P\rangle^\Lambda P(\Lambda)\, \overline{\langle P\rangle^\Lambda}$$

holds for every term P which is not a single variable. Thus the word $\langle P\rangle^{\mathcal{A}}$ contains the essential of the word $\langle P\rangle$. By construction the first character in $\langle P\rangle_u^v$ is $P(u)$, but of course this occurrence may disappear when reduction is operated.

Definition.- The term P is *good* if for every point u in Supp$P$ the word red$\langle P\rangle_u$ begins with $P(u)$.

Example.- Let Q be the term $a[b[c][b][b[c][d]]]$. This term is good, as shows the following array.

u	$\langle Q\rangle_u$	red$\langle Q\rangle_u$		
0	$abc\bar{b}bb\bar{c}\bar{b}bc\bar{b}db\bar{c}\bar{b}bc\bar{b}\bar{b}b\bar{c}\bar{b}\bar{a}$	abc	d	$\bar{c}\bar{b}\bar{a}$
1000	$bc\bar{b}bb\bar{c}\bar{b}bc\bar{b}db\bar{c}\bar{b}bc\bar{b}\bar{b}b\bar{c}\bar{b}\bar{a}$	bc	d	$\bar{c}\bar{b}\bar{a}$
1001	$c\bar{b}bb\bar{c}\bar{b}bc\bar{b}db\bar{c}\bar{b}bc\bar{b}\bar{b}b\bar{c}\bar{b}\bar{a}$	c	d	$\bar{c}\bar{b}\bar{a}$
101	$bb\bar{c}\bar{b}bc\bar{b}db\bar{c}\bar{b}bc\bar{b}\bar{b}b\bar{c}\bar{b}\bar{a}$	b	d	$\bar{c}\bar{b}\bar{a}$
1100	$bc\bar{b}db\bar{c}\bar{b}bc\bar{b}\bar{b}b\bar{c}\bar{b}\bar{a}$		$bc\bar{b}d$	$\bar{c}\bar{b}\bar{a}$
1101	$c\bar{b}db\bar{c}\bar{b}bc\bar{b}\bar{b}b\bar{c}\bar{b}\bar{a}$		$c\bar{b}d$	$\bar{c}\bar{b}\bar{a}$
111	$db\bar{c}\bar{b}bc\bar{b}\bar{b}b\bar{c}\bar{b}\bar{a}$		d	$\bar{c}\bar{b}\bar{a}$

Actually the example above seems to prove slightly more than the goodness of the word $\langle Q\rangle$: for every u in Supp$Q$ the occurrence of $Q(u)$ which begins the word red$\langle Q\rangle_u$ is precisely the one which comes from the original initial occurrence of $Q(u)$ in $\langle Q\rangle$. In other terms we have the more precise property that the words red$\langle Q\rangle'_u$ do not begin with $\overline{Q(u)}$, where $\langle Q\rangle'_u$ denotes the word obtained from $\langle Q\rangle_u$ by deleting the first character.

It is clear that any injective term P is good, since the word $\langle P\rangle$ is a reduced word. We make the following

Conjecture $(\mathcal{C})$.- *Any term which is $\approx$-equivalent to an injective term is good.*

Proposition 1.- *Assume that Q is an injective term, and Q^σ is good. Then the algorithm $(\mathcal{A})$ starting from any term P which is $\approx$-equivalent to Q^σ and from Q succeeds and returns the substitution σ.*

Proof. Let $y_1, \ldots, y_n$ be the variables of Q enumerated from the left to the right. Let w_j be the address of y_j in Q. By hypothesis the words $\langle P\rangle$ and $\langle Q^\sigma\rangle$ are $\equiv$-equivalent, hence they have the same reduced form. At the first step, we know that the word red$\langle Q^\sigma\rangle$ begins with the leftmost variable in Q^σ. So the first substitution σ_1 is the correct one: $y_1 \mapsto \sigma(y_1)$. Assume that the substitution σ_{j-1} constructed after $j-1$ steps is the restriction of σ to $\{y_1, \ldots, y_{j-1}\}$. At the j-th step, the word preceding the first occurrence of y_j is by construction

$$\mathrm{red}(\overline{\langle P\rangle}\langle Q^{\sigma_{j-1}}\rangle^{w_j}).$$

By hypothesis the word $\langle Q^{\sigma_{j-1}}\rangle^{w_j}$ coincides with $\langle Q^{\sigma}\rangle^{w_j}$, since the only variables which are involved in $\langle Q\rangle^{w_j}$ are $y_1, \ldots, y_{j-1}$. So the word above is also red($\overline{\langle Q^{\sigma}\rangle}\langle Q^{\sigma}\rangle^{w_j}$), which is red$\langle Q^{\sigma}\rangle_{w_j}$. Since Q^{σ} is assumed to be good, this word begins with $Q^{\sigma}(w_j)$, *i.e.* with $\sigma(y_j)$. Therefore its inverse ends with $\overline{\sigma(y_j)}$, and the algorithm guesses the correct value for the variable y_j. Thus the induction goes on. □

Proposition 2.- *Assume that Conjecture* ($\mathcal{C}$) *is true. Then the algorithm* ($\mathcal{A}$) *solves the Naming Problem for injective terms in exponential time, i.e. it gives the solution when it exists and fails when the solution does not exist.*

Proof. The fact that the algorithm ($\mathcal{A}$) works in exponential time (with respect to the size of the terms) is clear, since the length of the word $\langle Q\rangle$ is bounded by $2^{\mathrm{size}(Q)}$. By Proposition 1 we know that the algorithm succeeds for any pair of injective terms (P, Q) such that Q^{σ} is $\approx$-equivalent to P for some σ. It remains to show that the algorithm fails when no such σ exists. We use the Comparison Property (*c.f.* [5]) which, in the present framework, can be stated as follows. If no substitution σ exists satisfying $Q^{\sigma} \approx P$, then there must exist an integer $k \geq 1$, a sequence of injective terms $R_1, \ldots, R_k$ whose variables are disjoint from the variables occurring in P or Q and a substitution σ satisfying either $Q^{\sigma} \approx P[R_1]\ldots[R_k]$ or $(Q[R_1]\ldots[R_k])^{\sigma} \approx P$. In the first case, algorithm ($\mathcal{A}$) succeeds on the pair $(P[R_1]\ldots[R_k], Q)$. Let z be the leftmost variable in R_1. Let α be the word red($\overline{\langle P[R_1]\ldots[R_k]\rangle}\langle P\rangle$): this word cannot be empty, and its last character is $\bar{z}$. Let y the leftmost variable in Q such that $\sigma(y)$ is z. Since the algorithm works, at some step say j, the current word α_j has the form $\alpha y\beta$, and σ_j adds the value $y \mapsto z$. Let us consider the action of the algorithm ($\mathcal{A}$) on the pair (P, Q). The first $j-1$ steps are the same as above since the only part of the word $\overline{\langle P[R_1]\ldots[R_k]\rangle}$ used so far is the suffix $\overline{\langle P\rangle}$. It follows that the corresponding word α'_j is now $y\beta$, and the algorithm fails since the first occurrence of y cannot be cancelled. Assume now that $(Q[R_1]\ldots[R_k])^{\sigma}$ is equivalent to P. Then algorithm ($\mathcal{A}$) succeeds for the pair $(P, Q[R_1]\ldots[R_k])$. No obstruction will therefore occur when the algorithm is operated on the pair (P, Q), and one obtains the correct substitution namely the restriction of σ to the variables occurring in Q. But the final word obtained from the pair (P, Q) will be red($\overline{\langle P\rangle}Q^{\sigma}$). The word red($\overline{\langle Q^{\sigma}\rangle}\langle P\rangle$) is red($\overline{\langle Q^{\sigma}\rangle}\langle Q[R_1]\ldots[R_k]^{\sigma}\rangle$), which is also red($\langle Q[R_1]\ldots[R_k]^{\sigma}\rangle_{0^{k-1}10^i}$) where i is such that 0^i lies in the support of the term R_1. Since the term $Q[R_1]\ldots[R_k]^{\sigma}$ is $\approx$-equivalent to the injective term P, we know that this word begins with the variable $Q[R_1]\ldots[R_k](0^{k-1}10^i)$ (which is z), and therefore is not empty. Thus the algorithm fails for the pair (P, Q) in this case too. □

A natural way for proving the conjecture above is to find an inductive argument for the elementary steps of distribution.

Lemma 3.- *Assume that the transformation Ω_w maps the term P to the term Q, and u lies in the support of Q.*

i) If u and w are $\sqsubset$-incomparable or $w11 \sqsubset u$ holds, then $\langle Q\rangle_u \equiv \langle P\rangle_u$ holds.
ii) If u is $w00u'$, then $\langle Q\rangle_u \equiv \langle P\rangle_{w0u'}$ holds.
iii) If u is $w01u'$, then $\langle Q\rangle_u \equiv \langle P\rangle_{w10u'}$ holds.
iv) If u is $w10u'$, then $\langle Q\rangle_u \equiv \langle P\rangle_{w0u'}^{w10^j}\langle P\rangle_{w110^k}$ holds, where j and k are such that $w10^j$ and $w110^k$ belong to the support of P.

The proof is easy. If P is known to be good, it suffices in order to prove the goodness of the image of P under Ω_w to study cancellation in the words $\langle Q\rangle_u \equiv \langle P\rangle_{w0u'}^{w10^j}\langle P\rangle_{w110^k}$. Such a word is obtained from $\langle P\rangle_{w0u}$ by deleting the subword $\langle P\rangle_{w10^j}^{w110^k}$. By hypothesis we know that the word red$\langle P\rangle_{w0u}$ begins with $P(w0u)$. The point is to control the new cancellations which can be induced by the deletion of the subword above.

4. Goodness of derived terms.

By Lemma 3.3 it is clear that any term having a good extension must be good itself. Since $\approx$-equivalent terms must have a common extension, it suffices to show that any extension of an injective term is good for proving Conjecture $(\mathcal{C})$. In [2] we have constructed for every term P a sequence of terms

$$P, \partial P, \partial^2 P, etc\ldots$$

such that every term $\approx$-equivalent to P admits as an extension every term $\partial^n P$ for n large enough. Therefore we have

Lemma 1.- *Assume that, if the term P is injective, all terms $\partial^n P$ are good. Then Conjecture$(\mathcal{C})$ is true.*

We shall establish a partial result in this direction. First we recall the construction of the term derivation ∂.

Definition.- i) For P, Q in $\mathcal{T}_\Sigma$, the term $P[\![Q]\!]$ is defined inductively on Q by

$$P[\![Q]\!] = \begin{cases} P[Q] & \text{if } Q \text{ is a single variable,} \\ P[\![Q_0]\!][P[\![Q_1]\!]] & \text{if } Q \text{ is } Q_0[Q_1]. \end{cases}$$

ii) For P is $\mathcal{T}_\Sigma$, the term ∂P is defined inductively by

$$\partial P = \begin{cases} P & \text{if } P \text{ is a single variable,} \\ \partial(P_0)[\![\partial(P_1)]\!] & \text{if } P \text{ is } P_0[P_1]. \end{cases}$$

The term $P[\![Q]\!]$ is obtained from the term Q by replacing every variable y of Q by the term $P[y]$, *i.e.* by distributing P everywhere in Q.

Lemma 2.- *Let P, Q be arbitrary terms. Then the support of the term $P[\![Q]\!]$ is*

$$(\mathrm{Supp}Q)0(\mathrm{Supp}P)\cup(\mathrm{Supp}Q)1.$$

The equivalences

$$\langle P[\![Q]\!]\rangle_{v0u}\equiv\langle P\rangle_u\langle Q\rangle_v\overline{\langle P\rangle}\text{ and }\langle P[\![Q]\!]\rangle_{v1}\equiv\langle Q\rangle_v\overline{\langle P\rangle}$$

hold for u in Supp$P$ *and v in* Supp$Q$.

Proof. If Q is a single variable, $P[\![Q]\!]$ is equal to $P[Q]$. Hence the support of $P[\![Q]\!]$ is $0(\mathrm{Supp}P)\cup\{1\}$. For u in Supp$P$ one has

$$\langle P[\![Q]\!]\rangle_{0u}=\langle P[Q]\rangle_{0u}=\langle P\rangle_u Q\overline{\langle P\rangle}=\langle P\rangle_u\langle Q\rangle_\Lambda\overline{\langle P\rangle}.$$

If Q is $Q_0[Q_1]$, one has

$$\begin{aligned}\mathrm{Supp}(P[\![Q]\!])&=0(\mathrm{Supp}(P[\![Q_0]\!])\cup 1(\mathrm{Supp}(P[\![Q_1]\!])\\&=0((\mathrm{Supp}Q_0)0(\mathrm{Supp}P)\cup(\mathrm{Supp}Q_0)1)\cup\\&\qquad\qquad 1((\mathrm{Supp}Q_1)0(\mathrm{Supp}P)\cup(\mathrm{Supp}Q_1)1)\\&=(0(\mathrm{Supp}Q_0)\cup 1(\mathrm{Supp}Q_1))0(\mathrm{Supp}P)\cup\\&\qquad\qquad(0(\mathrm{Supp}Q_0)\cup 1(\mathrm{Supp}Q_1))0(\mathrm{Supp}P)1\\&=(\mathrm{Supp}Q)0(\mathrm{Supp}P)\cup(\mathrm{Supp}Q)1.\end{aligned}$$

Assume that u belongs to Supp$P$ and v belongs to Supp$Q$. Write $v=ev'$ with $e=0$ or $e=1$. Assume first $e=0$. One has

$$\begin{aligned}\langle P[\![Q]\!]\rangle_{v0u}&=\langle P[\![Q_0]\!][P[\![Q_1]\!]]\rangle_{0v'0u}\\&=\langle P[\![Q_0]\!]\rangle_{v'0u}\langle P[\![Q_1]\!]\rangle\overline{\langle P[\![Q_0]\!]\rangle}\\&=\langle P\rangle_u\langle Q_0\rangle_{v'}\overline{\langle P\rangle}\langle P\rangle\langle Q_1\rangle\overline{\langle P\rangle}\langle P\rangle\overline{\langle Q_0\rangle}\,\overline{\langle P\rangle}\\&\equiv\langle P\rangle_u\langle Q_0\rangle_{v'}\langle Q_1\rangle\overline{\langle Q_0\rangle}\,\overline{\langle P\rangle}\\&=\langle P\rangle_u\langle Q\rangle_{0v'}\overline{\langle P\rangle}\\\langle P[\![Q]\!]\rangle_{v1}&=\langle P[\![Q_0]\!][P[\![Q_1]\!]]\rangle_{0v'1}\\&=\langle P[\![Q_0]\!]\rangle_{v'1}\langle P[\![Q_1]\!]\rangle\overline{\langle P[\![Q_0]\!]\rangle}\\&=\langle Q_0\rangle_{v'}\overline{\langle P\rangle}\langle P\rangle\langle Q_1\rangle\overline{\langle P\rangle}\langle P\rangle\overline{\langle Q_0\rangle}\,\overline{\langle P\rangle}\\&\equiv\langle Q_0\rangle_{v'}\langle Q_1\rangle\overline{\langle Q_0\rangle}\,\overline{\langle P\rangle}\\&=\langle Q\rangle_{0v'}\overline{\langle P\rangle}\end{aligned}$$

Similarly assuming $e=1$ one obtains

$$\begin{aligned}\langle P[Q]\rangle_{v0u} &= \langle P[Q_0][P[Q_1]]\rangle_{1v'0u} \\ &= \langle P[Q_1]\rangle_{v'0u}\overline{\langle P[Q_0]\rangle} \\ &= \langle P\rangle_u\langle Q_1\rangle_{v'}\overline{\langle P\rangle}\langle P\rangle\overline{\langle Q_0\rangle}\,\overline{\langle P\rangle} \\ &\equiv \langle P\rangle_u\langle Q_1\rangle_{v'}\overline{\langle Q_0\rangle}\,\overline{\langle P\rangle} \\ &= \langle P\rangle_u\langle Q\rangle_{1v'}\overline{\langle P\rangle}\end{aligned} \qquad \begin{aligned}\langle P[Q]\rangle_{v1} &= \langle P[Q_0][P[Q_1]]\rangle_{1v'1} \\ &= \langle P[Q_1]\rangle_{v'1}\overline{\langle P[Q_0]\rangle} \\ &= \langle Q_1\rangle_{v'}\langle P\rangle\overline{\langle Q_0\rangle}\,\overline{\langle P\rangle} \\ &\equiv \langle Q_1\rangle_{v'}\overline{\langle Q_0\rangle}\,\overline{\langle P\rangle} \\ &= \langle Q\rangle_{1v'}\overline{\langle P\rangle}\end{aligned}$$

which finishes the proof. □

Proposition 3.- *If the term P is injective, the term ∂P is good.*

Proof. Use induction on the size of the term P. The result is obvious if P is a single variable. Assume that P is $P_0[P_1]$, and that ∂P_0 and ∂P_1 are good by induction hypothesis. Then the variables in ∂P_0 and ∂P_1 are disjoint (because P is injective). Lemma 1 yields for u in $\mathrm{Supp}\partial P_0$ and v in $\mathrm{Supp}\partial P_1$

$$\begin{aligned}\mathrm{red}\langle\partial P\rangle_{v0u} &= \mathrm{red}\langle\partial P_0\rangle_u\mathrm{red}\langle\partial P_1\rangle_v\mathrm{red}\overline{\langle\partial P_0\rangle} \\ \mathrm{red}\langle\partial P\rangle_{v1} &= \mathrm{red}\langle\partial P_1\rangle_v\mathrm{red}\overline{\langle\partial P_0\rangle}\end{aligned}$$

and because the variables in ∂P_0 and ∂P_1 are disjoint no cancellation may happen between the words $\mathrm{red}\langle\partial P_0\rangle_u$ and $\mathrm{red}\langle\partial P_1\rangle_v$. So the goodness of ∂P_0 and ∂P_1 implies that $\mathrm{red}\langle\partial P\rangle_{v0u}$ begins like $\mathrm{red}\langle\partial P_0\rangle_u$, *i.e.* begins with $\partial P_0(u)$, which is $\partial P(v0u)$, and that $\mathrm{red}\langle\partial P\rangle_{v1}$ begins like $\mathrm{red}\langle\partial P_1\rangle_v$, *i.e.* begins with $\partial P_1(v)$, which is $\partial P(v1)$. Hence ∂P is good. □

Using the notion of a *simple* extension, which can be defined by the property that Q is a simple extension of P if Q is an extension of P and ∂P is an extension of Q, we may conclude that any simple extension of an injective term is good. So in particular the algorithm $(\mathcal{A})$ works for simple extensions. With more care one may give a precise description of the words $\langle\partial^2 P\rangle$ for P an injective term, and obtain a proof of the goodness of the second derivative of an injective term. Unfortunately this method seems hard to iterate further.

5. Skew conjugacy.

We finish this paper by introducing a "twisted" version of distribution and conjugacy which leads to a variant of the algorithm $(\mathcal{A})$. Although this new algorithm seems to be even more complicated than the original one, we shall see that it always works by using an auxiliary ordering.

Left distributivity is associated with the operators Ω_w, which are translated copies of the unique operator Ω_Λ defined by

$$\Omega_\Lambda : P[Q[R]] \mapsto P[Q][P[R]].$$

The idea is to replace "true" left distributivity by the similar equivalence associated with a modified operator

$$\Omega^\sharp_\Lambda : P[Q[R]] \mapsto P[Q][\tau_Q(P)[R]],$$

where the term $\tau_Q(P)$ is obtained from the term P by replacing any variable x occurring in P by a new variable depending on x and on the term Q. We are interested in the case where the geometry of the operator $\Omega^\sharp_\Lambda$ and its translated copies $\Omega^\sharp_w$ is the same as the geometry of the operators Ω_w in order to be able to project the results obtained for the modified operators into results for the initial operators. This happens when $\tau_Q(x)$ only depends on the rightmost variable in Q, and moreover the operation $(y, x) \mapsto \tau_y(x)$ is itself a left distributive operation on the set of all variables. We shall write ${}^y x$ for the product of y and x for this operation.

Example.- Let P be the term $a[b[c[d]]]$. The image of P under $\Omega^\sharp_\Lambda$ is the term $a[b][{}^b a[c[d]]]$. Similarly the image of P under $\Omega^\sharp_1$ is the term $a[b[c][{}^c b[d]]]$.

Observe that, if the operation is the trivial (left distributive) operation defined by ${}^y x = x$, then the modified operators $\Omega^\sharp_w$ coincide with the standard operators Ω_w. In the sequel we shall consider the "opposite" case where the operation is the bracket of a free left distributive magma. Our notations will be the following ones. For any set Σ we let $\Sigma^\sharp$ be the free left distributive magma generated by Σ (so that $\Sigma^\sharp$ is isomorphic to the quotient $T_\Sigma/\approx$). We consider terms with variables in $\Sigma^\sharp$, *i.e.* in $T_{\Sigma^\sharp}$. Observe that there exists a welldefined projection

$$\pi : \Sigma^\sharp \to \Sigma$$

which is the identity on Σ and always satisfies $\pi({}^y x) = \pi(x)$. We extend π to terms coordinatewise. Finally we we say that Q is a $\sharp$-extension of P if Q is obtained from P by applying a composition of (finitely many) operators $\Omega^\sharp_w$, and we denote by $\approx^\sharp$ the congruence on $T_{\Sigma^\sharp}$ generated by all operators $\Omega^\sharp_w$.

Lemma 1.- *i) Assume that P, Q are $\approx^\sharp$-equivalent terms in $T_{\Sigma^\sharp}$. Then $\pi(P)$ and $\pi(Q)$ are $\approx$-equivalent terms.*

ii) Conversely if Q is an extension of P, there exists a unique $\sharp$-extension $Q^\sharp$ of P such that Q is $\pi(Q^\sharp)$.

Proof. With the above notations, we always have

$$\pi(\tau_Q(P)) = \pi(P),$$

since $\pi({}^y x)$ is equal to $\pi(x)$. It follows that $\pi(Q)$ is always an extension of $\pi(P)$ whenever Q is a $\sharp$-extension of P.

Now assume that Q is an extension of P, *i.e.* Q is obtained from P by successively applying the operators $\Omega_{w_1}, \ldots, \Omega_{w_n}$. Let $Q^\sharp$ be the image of P under the composition of $\Omega^\sharp_{w_1}, \ldots, \Omega^\sharp_{w_n}$. First of all the terms Q and $Q^\sharp$ have the same support, and clearly Q is $\pi(Q^\sharp)$. The main point is to verify that the term $Q^\sharp$ does not depend on the choice of the points $w_1, \ldots, w_n$. To this end one has to know all relations between the compositions of operators Ω_w. If this is known it then suffices to verify that the corresponding relations hold for the operators $\Omega^\sharp$. The first step follows from the general study of left distributivity (see [5]), the second one is the place where the necessity of using a left distributive operation in the definition of the modified operators appears. We shall not give the details here. □

Assume that we are given two injective terms P, Q in $\mathcal{T}_\Sigma$, we may consider the $\sharp$-Naming Problem which consists in finding a substitution

$$\sigma : \Sigma \to \Sigma^\sharp$$

such that the term Q^σ is $\approx^\sharp$-equivalent to the term P if it exists. By Lemma 1 above, any solution to this question will automatically yield a solution to the corresponding Naming Problem by using the projection π. Conversely any solution to the Restricted Naming Problem can be lifted in a unique way to a solution for the associated $\sharp$-Problem. The latter one seems more complicated since, starting with the support of a term, we have not only to guess the values of the "short" variables in Σ but also the values of the "long" variables in $\Sigma^\sharp$.

In order to solve this $\sharp$-Naming Problem, we introduce a skew version of conjugacy. We write $\mathcal{W}_{\Sigma^\sharp}$ for the free monoid generated by $\Sigma^\sharp$ and a disjoint copy $\overline{\Sigma^\sharp}$ of $\Sigma^\sharp$, and $\mathcal{G}_{\Sigma^\sharp}$ for the free group generated by $\Sigma^\sharp$.

Definition.- For P in $\mathcal{T}_{\Sigma^\sharp}$, $\langle P\rangle^\sharp$ is the word in $\mathcal{W}_{\Sigma^\sharp}$ inductively defined by

$$\langle P\rangle^\sharp = \begin{cases} P & \text{if } P \text{ lies in } \Sigma^\sharp, \\ \langle P_0\rangle^\sharp \langle P_1\rangle^\sharp \overline{\langle {}^y P_0\rangle^\sharp} & \text{if } P \text{ is } P_0[P_1] \text{ and } y \text{ is the rightmost variable in } P_1. \end{cases}$$

Example.- Assume that Q is the term $A[B][C][D[E]][F]]]$. Then the word $\langle Q\rangle^\sharp$ is

$$A\,B\,\overline{{}^B A}\,C\,{}^C({}^B A)\,\overline{{}^C B}\,\overline{{}^C A}\,D\,E\,\overline{{}^E D}\,F\,{}^F({}^E D)\,\overline{{}^F E}\,\overline{{}^F D}$$
$${}^F({}^C A)\,{}^F({}^C B)\,\overline{{}^F({}^C({}^B A)))}\,\overline{{}^F C}\,{}^F({}^B A)\,\overline{{}^F B}\,\overline{{}^F A},$$

whose projection under π is the following word $\langle Q\rangle$ (we assume that $A, \ldots, F$ belong to Σ)

$$A\,B\,\overline{A}\,C\,A\,\overline{B}\,\overline{A}\,D\,E\,\overline{D}\,F\,D\,\overline{E}\,\overline{D}\,A\,B\,\overline{A}\,\overline{C}\,A\,\overline{B}\,\overline{A}.$$

Lemma 2.- *If P, Q are in $T_{\Sigma^\natural}$ and $P \approx^\natural Q$ holds, then the words $\langle P\rangle^\natural$ and $\langle Q\rangle^\natural$ are $\equiv$-equivalent.*

Proof. It suffices to show the equivalence

$$\langle P[Q[R]]\rangle^\natural \equiv \langle P[Q][{}^{y}P[R]]\rangle^\natural$$

where y is the rightmost variable in Q. Let z be the rightmost variable in R. We have

$$\begin{aligned}
\langle P[Q[R]]\rangle^\natural &= \langle P\rangle^\natural\langle Q\rangle^\natural\langle R\rangle^\natural\overline{\langle {}^{z}Q\rangle^\natural}\,\overline{\langle {}^{z}P\rangle^\natural}\\
\langle P[Q][{}^{y}P[R]]\rangle^\natural &= \langle P\rangle^\natural\langle Q\rangle^\natural\overline{\langle {}^{y}P\rangle^\natural}\langle {}^{y}P\rangle^\natural\langle R\rangle^\natural\overline{\langle {}^{z}({}^{y}P)\rangle^\natural}\langle {}^{z}({}^{y}P)\rangle^\natural\overline{\langle {}^{z}Q\rangle^\natural}\,\overline{\langle {}^{z}P\rangle^\natural}\\
&\equiv \langle P\rangle^\natural\langle Q\rangle^\natural\langle R\rangle^\natural\overline{\langle {}^{z}Q\rangle^\natural}\,\overline{\langle {}^{z}P\rangle^\natural}
\end{aligned}$$

which gives the result. □

When using the algorithm $(\mathcal{A})$ with the words $\langle P\rangle^\natural$ instead of $\langle P\rangle$, we obtain a new method which we shall denote by $(\mathcal{A}^\natural)$.

Example.- Let P be the term $a[b[c]]$ and Q be $A[B][C][D[E]][F]]]$ as above. Applying $(\mathcal{A}^\natural)$ to the pair (P, Q) leads to the following steps. For simplicity we only write the first half of the corresponding words.

$$\begin{array}{lllllllllllllll}
 & \ldots\overline{c} & \overline{b} & \overline{a} & A & B & \overline{{}^{B}A} & C & {}^{C}({}^{B}A) & \overline{{}^{C}B} & \overline{{}^{C}A} & D & E & \overline{{}^{E}D} & F\ldots\\
A \mapsto a & \ldots\overline{c} & \overline{b} & & & B & \overline{{}^{B}a} & C & {}^{C}({}^{B}a) & \overline{{}^{C}B} & \overline{{}^{C}a} & D & E & \overline{{}^{E}D} & F\ldots\\
B \mapsto b & \ldots\overline{c} & & & & & \overline{{}^{b}a} & C & {}^{C}({}^{b}a) & \overline{{}^{C}b} & \overline{{}^{C}a} & D & E & \overline{{}^{E}D} & F\ldots\\
C \mapsto {}^{b}a & \ldots\overline{c} & & & & & & & {}^{{}^{b}a}({}^{b}a) & \overline{{}^{{}^{b}a}b} & \overline{{}^{{}^{b}a}a} & D & E & \overline{{}^{E}D} & F\ldots\\
D \mapsto {}^{{}^{b}a}a & \ldots\overline{c} & & & & & & & {}^{{}^{b}a}({}^{b}a) & \overline{{}^{{}^{b}a}b} & & & E & \overline{{}^{E}{}^{{}^{b}a}a} & F\ldots\\
E \mapsto {}^{{}^{b}a}b & \ldots\overline{c} & & & & & & & {}^{{}^{b}a}({}^{b}a) & & & & & & F\ldots\\
F \mapsto c & & & & \varepsilon & & & & & & & & & &
\end{array}$$

Remark.- The algorithm $(\mathcal{A}^\natural)$ is a "false" algorithm in so far as it requires another algorithm for recognizing equality in $\Sigma^\natural$ in order to correctly control the cancellations in $\mathcal{W}_{\Sigma^\natural}$. For instance in the example above one has to recognize, just before the last step, that the character $\overline{{}^{E}D}$, which has become $\overline{{}^{{}^{{}^{b}a}b}({}^{{}^{b}a}a)}$, vanishes with the character ${}^{{}^{b}a}({}^{b}a)$, which is true since the operation ${}^{y}x$ is left distributive. The elements of $\Sigma^\natural$ are themselves represented by words constructed from Σ and the operation ${}^{y}x$. There exists unique normal forms for $\Sigma^\natural$ (see [7] or [8]), which can be used in this case.

But finding the normal form of an arbitrary term is as difficult as solving the Word Problem for the congruence $\approx$ (since two terms are $\approx$-equivalent if and only if their normal forms coincide). This problem is known to be decidable (see [5]), but the Naming Problem is a particular case of it, and it looks strange to use the solution to the more general problem as a preliminary tool for constructing a solution to the particular case.

The main tool we shall use now is the existence of linear orderings on $\Sigma^{\natural}$ established in [4]. Fix any linear ordering $<$ on Σ. Then there exists a linear ordering $<^{\natural}$ on $\Sigma^{\natural}$ which is compatible with left translations and satisfies $y <^{\natural} {}^{y}x$ for every x, y in $\Sigma^{\natural}$. The ordering $<^{\natural}$ is induced on $\Sigma^{\natural}$ by the lexicographical extension $<_{\mathrm{Lex}}$ of $<$ to $\mathcal{T}_{\Sigma}$ when right Polish notation is used for terms. Thus $P <_{\mathrm{Lex}} Q$ holds if either P (in right Polish notation) is a prefix of Q or if P is $\alpha a\beta$ and Q is $\alpha b\gamma$ with a, b in Σ satisfying $a < b$. We write $x \ll y$ in $\Sigma^{\natural}$ if x and y are the classes (under $\approx$) of terms P, Q satisfying the second condition above.

Assume that P is any term in $\mathcal{T}_{\Sigma^{\natural}}$, and that w is a point in Supp$P$ which contains at least one 1 (so w is not the leftmost point in Supp$P$). Then w can be written in a unique way as $u10^i$. The unique points in Supp$P$ of the form $u0^j$ and $u01^k$ are denoted respectively by $\varphi_P(w)$ and $\psi_P(w)$ (they may coincide).

Definition.- A term P in $\mathcal{T}_{\Sigma^{\natural}}$ is *strongly increasing* if for every point w in Supp$P$ (except the first one) at least one of $P(\psi_P(w)) \ll P(w)$ and $P(w) =^{P(\psi_P(w))} P(\varphi_P(w))$ holds.

Strongly increasing terms are clearly increasing in the sense that their variables make an increasing sequence when enumerated from the left to the right since for every w in Supp$P$ the point $\psi_P(w)$ is the immediate predecessor of w in Supp$P$. Also an injective term with variables in Σ is strongly increasing whenever the ordering $<$ on Σ extends the left-right ordering of its variables.

Lemma 3.- *Any $\natural$-extension of a strongly increasing term is a strongly increasing term.*

Proof. Assume that P is strongly increasing and that Q is obtained from P by applying the operator $\Omega^{\natural}_w$. The point is to study the values of the new variables, *i.e.* of the variables $Q(w10u)$. By construction $Q(w10u)$ is ${}^{Q(w10)}Q(w0u)$. Left translation associated to $Q(w10)$ preserves the "inner" conditions for the points $Q(w0u)$: $Q(w0u') \ll Q(w0u)$ implies ${}^{Q(w10)}Q(w0u') \ll^{Q(w10)} Q(w0u)$, and $Q(w0u) =^{Q(w0u')} Q(w0u'')$ implies (by left distributivity) ${}^{Q(w10)}Q(w0u) =^{{}^{Q(w10)}Q(w0u')} ({}^{Q(w10)}Q(w0u''))$. For the extremal points we always have (if $w0^i$ belongs to Supp$P$)

$$Q(w10^i) =^{Q(w10)} Q(w0^i) =^{Q(\psi_Q(w10^i))} Q(\varphi_Q(w10^i)).$$

Assume that $w0^i$, $w01^j$, $w10^k$, $w101^\ell$ and $w110^m$ belong to Supp$P$, and let x', x, y', y and z the corresponding variables in P. If $y \ll z$ holds, then ${}^y x \ll z$ also holds by construction of the lexicographical ordering. Now assume $z = {}^y y'$. If $x \ll y'$ holds, then ${}^y x \ll {}^y y'$, *i.e.* ${}^y x \ll z$, follows. And if $y' = {}^x x'$ holds, we obtain by left distributivity $z = {}^y y' = {}^y({}^x x') = {}^{{}^y x}({}^y x')$. So in any case all conditions are fulfilled. □

We are now able to establish the following counterpart of Conjecture ($\mathcal{C}$). We shall say that a term Q is $\sharp$-good if the word red$\langle Q\rangle^{\sharp}_u$ begins with $Q(u)$ for u in the support of Q.

Proposition 4.- *Every term which is $\approx^\sharp$-equivalent to an injective term is $\sharp$-good.*

Proof. Assume that P is an injective term in $\mathcal{T}_\Sigma$. Fix any linear ordering $<$ on Σ such that the variables of P form an increasing sequence when enumerated from the left to the right. Then by the above lemma every $\sharp$-extension of P will be a strongly increasing term. We claim that any (strongly) increasing term must be $\sharp$-good. This will show that any $\sharp$-extension of an injective term is $\sharp$-good, and (using formulas similar to the formulas in Lemma 3.3 for the case of a $\sharp$-extension) this implies that any term which is $\approx^\sharp$-equivalent to an injevtive term is $\sharp$-good. So assume that Q is (strongly) increasing. Let w be a point in Supp$Q$, and let y be $Q(w)$. Then the word $\langle Q\rangle^{\sharp}_w$ begins with y, and all subsequent characters must be bigger than y. Indeed let w' be the next point in Supp$Q$. By definition all characters in $\langle Q\rangle^{\sharp w'}_w$ have the form ${}^y x$ for some x, and therefore are bigger than y. By hypothesis y' is bigger than y, and by induction every character after the first occurrence of y' is bigger than y', hence than y. So no character in $\langle Q\rangle^{\sharp}_w$ may attack the first occurrence of y, and the word red$\langle Q\rangle^{\sharp}_w$ begins with y. □

As in Section 3 we can deduce the following

Corollary 5.- *The algorithm ($\mathcal{A}^\sharp$) solves the $\sharp$-Naming Problem for injective terms.*

The examples suggest that the only "meaningful" cancellations (*i.e.* the ones involving first occurrences of variables in Q) happening in the words $\langle\pi(Q)\rangle_w$ are the ones corresponding to cancellations in the words $\langle Q\rangle^{\sharp}_w$. If this is true, Conjecture ($\mathcal{C}$) is true. But no general argument is known for the moment.

References.

[1] R. H. Crowell & R. H. Fox Knot Theory, Blaisdell Publ. Comp., (1963)

[2] P. Dehornoy, *Free distributive groupoids*, Journal of Pure and Applied Algebra, **61** (1989) 123–146.

[3] —, *Problème de mots dans les gerbes libres*, Theor. Comp. Sc., **94** (1992) 199–213.
[4] —, *A canonical ordering for free distributive magmas*, Proc. of the AMS, *to appear*.
[5] —, *Braid Groups and Left Distributive Operations*, preprint (1992).
[6] G. HUET & D. OPPEN, *Equations and Rewrite Rules: a survey*, in R. Book, ed., Formal Languages: Perspectives and Open Problems, Academic Press (1980).
[7] R. LAVER, *The left distributive law and the freeness of an algebra of elementary embeddings*, Advances in Mathematics, **91-2** (1992) 209–231.
[8] —, *A division algorithm for the free left distributive algebra*, Proc. Helsinki 1990 ASL Meeting *to appear*.

Mathématiques, Université, 14 032 Caen, France
dehornoy@geocub.greco-prog.fr

A case of termination for associative unification

Patrice Enjalbert - Françoise Clerin-Debart
Université de Caen 14032 Caen Cedex France

Introduction

In studying automated deduction for modal logics [Auffray, Enjalbert 89] [Debart, Enjalbert, Lescot 90] by equational means we got into problems of associative unification. It happens that we were able to design a terminating (recursive) unification algorithm in our particular case, thanks to a certain property of the considered terms, known as Unique Prefix Property (and very close to Prefix Stability of [Ohlbach 88]). But the proof was very long and intricate, and associativity was mixed with other features of the typed equational theories we consider for modal logic. In order to simplify the proof and to change the design of the algorithm for a rule-based algorithm, we tried to extract from our results what is strictly relevant of associative unification.

We define a general "Unique Prefix Property" for sets of terms in a first order logic with an associative symbol. We prove that such a set of terms has a finite set of unifiers and that in this case Plotkin's algorithm [Plotkin 72] terminates. The proof uses a tool which seems specially well-fitted for this class of terms: the tree of prefixes. We show that each step of the unification process corresponds to some operation on this tree which can be repeated only a finite number of times.

An extension of this result to a class of rule-based algorithms, for associative and unitary unification and some order-sorted logics with over-loaded symbols, will also be shortly presented.

1.Preliminaries

In all the following we use a first-order language, with a set of functions symbols F. The set F contains a symbol + which is associative.

Let V denotes the set of variables and Var(t) the set of variables occurring in the term t.

In the following, $t\downarrow$ is the right normal form of t w.r.t. the associativity axiom, (x+y)+z is rewritten x+(y+z). We omit the brackets in normalised terms.

Recall that in a standard first-order logic, two unifiable terms have a unique m.g.u. (up to a renaming). This fact is no longer true in an associative theory: a+x and x+a have infinitely many incomparable unifiers, {x/a}, {x/a+a }, {x/a+a+a } ....[Fages Huet 86].

2. Plotkin's Algorithm

In [Plotkin 72] a non-deterministic algorithm for associative unification is presented.

Given two terms t_0 et t'_0, each step computes non deterministically, a triplet (t,t', σ) where t and t' are terms (t_0 et t'_0 partially unified) and σ a substitution. A successful computation ends with t=t' and then σ is a unifier of t and t'. The set of successful computations provides a complete sets of unifiers. Of course, in the general case, we can have infinitely many successes or failures

Let μ a substitution, and $\mu\downarrow$ the mapping such that $\mu\downarrow t = (\mu t)\downarrow$

Plotkin's Associative Unification Algorithm

<u>Begin</u>

$\sigma := \emptyset$;

<u>while</u> $t\downarrow \neq t'\downarrow$

find (u,u') the disagreement pair of $t\downarrow$ et $t'\downarrow$;

<u>Case of</u>

u , u' both non variables :	stop with failure
u' variable , u non variable:	exchange t, t'
u et u' both variables:	let μ_1={u/u'}, μ_2={u/u'+u}, μ_3= {u'/u+u'}
u variable, u' non variable :	<u>if</u> u is a subterm of u' <u>then</u> stop with failure
	<u>else</u> let μ_1= {u/u'}, μ_2={u/u'+u}

<u>end case;</u>

In a non-deterministic way:

<u>Choose</u> i; $\sigma := (\sigma\,\mu_i)\downarrow$; $t := \mu_i\downarrow t$; $t' := \mu'_i\downarrow t$

<u>end while</u>

<u>end</u>

First we shall establish properties of the set of subterms of a term. Then we define the **Unique Prefix Property** and finally we prove termination of Plotkin's algorithm for terms with the Unique Prefix Property.

3. Prefixes of a set of terms

3.1. Definitions and notations

Definition 1: We call **additive** a term like t_1+t_2, (the main functionnal symbol is +) and **non-additive** a term which is a variable or a term with a main fonctionnal symbol different from +.

Notations

Given a term t, we note $t^\bullet$ the list :

- $t^\bullet = (t\downarrow)$ if t is a non-additive term.
- $t^\bullet = (t_1, t_2, ..., t_n)$ if $t\downarrow = t_1 + t_2 + ... + t_n$ $(n>0)$ (where the t_i's are non-additive)

We denote by ϖ **the empty list** and by $\bullet$ **the concatenation of lists** . If p is a list, **$|p|$ denotes the length of p.**

Definition 2: A **prefix** is a (possibly empty) list of non-additive normalised terms .

Remark that if $p = (t_1,...,t_n)$ is a prefix, then $(t_1+...+t_n)^\bullet = p$.

Definition 3: Let σ be a substitution, $\sigma^\bullet$ maps a prefix p onto a prefix $\sigma^\bullet(p)$ such that:

- $\sigma^\bullet(\varpi) = \varpi$
- if p has only one element t_1(variable or not) $\sigma^\bullet(p) = (\sigma\downarrow t_1)^\bullet$
- if $p=(t_1...t_n)$ then $\sigma^\bullet(p) = (\sigma(t_1+t_2+....+t_n))^\bullet$.

In particular:

- For each term t $\sigma^\bullet(t^\bullet) = (\sigma t)^\bullet$
- if $p=(t_1,...,t_n)$ then $\sigma^\bullet(p)$ is the concatenation of all the lists $\sigma^\bullet(t_i)$.

3.2. Prefixes of a set of terms

Let t a term, we note O(t) the set of all occurrences in t. If u is in O(t), $t|_u$ denotes the subterm of t at occurrence u.

We call **non-additive** an occurrence u in O(t) such that $t|_u$ is a non additive term. We note $O^-(t)$ the set of all non-additive occurrences in t and $O^+(t) = O - O^-(t)$.

Definition 4: Given a term t, $\Pi(t)$ is the set of prefixes defined by:

$$\Pi(t) = \{t^\bullet\} \cup \{ u^\bullet / u = t|_{o.i},\ o \in O^-(t),\ o.i \in O(t)\}$$

We easily extend this definition to a set of terms $T=\{t_1,\dots,t_n\}$ by $\Pi(T) = \bigcup_i(\Pi(t_i))$.

Example 1: $T=\{f(a+g(b+x))\ ;\ a + g(a+b)\}$

$$\Pi(T) = \{\ (f(a+g(b+x))\ ), (a,g(b+x)), (b,x), (a,g(a+b)), (a,b)\}$$

The following properties are obvious:

- if t is a constant or a variable: $\Pi(t) = \{(t)\}$
- if t is $f(t_1,..,t_n)$ then $\Pi(t) = \{(t)\} \cup (\bigcup_{1\leq i\leq n} \Pi(t_i))$
- if $t = t_1+\dots+t_n$ is in normal form, then $\Pi(t) = \{(t)^\bullet\} \cup (\bigcup_{1\leq i\leq n} [\Pi(t_i) - \{(t_i)\}\ ])$.

If v is a subterm of t, then $\Pi(v) - \{(v)^\bullet\} \subset \Pi(t)$.

In particular:

- if u is in $O^-(t)$ $\Pi(t|_{u.i}) \subset \Pi(t)$. (for u.i in O(t))
- if u is in $O^+(t)$ $\Pi(t|_{u.i}) - \{(t|_{u.i})\} \subset \Pi(t)$ (for i=1,2).

3.3. Prefixes of an instance of a set of terms

Proposition 1: If σ is a substitution then $\sigma^\bullet\Pi(T) \subset \Pi(\sigma T)$.

More, if $\sigma=\{x/u\}$ or $\{x/u+x\}$ and if u is a subterm of T whith no occurrence of x then $\sigma^\bullet\Pi(T) = \Pi(\sigma T)$.

First, we give an example:

Example 2: Consider the set T of example 1 again:

let $\sigma=\{x/a+g(a+c)\}$

$\sigma T=\{f(a+g(b+a+g(a+c))\ ;\ a + g(a+b)\}$

$\sigma^\bullet(\Pi(T))=\{(f(a+g(b+a\ \ +g(a+c)))), (a,g(b+a+g(a+c))), (b,a,g(a+c)), (a,g(a+b)), (a,b)\}$

$\Pi(\sigma T) =\{\ f(a+g(b+a+g(a+c)\ )), (a,g(b+a+g(a+c)), (b,a,g(a+c)), \mathbf{(a,c)},$
$(a,g(a+b)), (a,b)\}$

We have $\sigma^\bullet\Pi(T) \subset \Pi(\sigma T)$, but a+c is not a sub-term of T and (a,c) is not in $\sigma^\bullet\Pi(T)$

Proof:

The first result is an immediate consequence of the properties stated in 3.2.

We have to prove that if $\sigma=\{x/u\}$ or $\{x/u+x\}$ and u is a subterm of T then $\Pi(\sigma T) \subset \sigma^\bullet\Pi(T)$.

For any subterm u of a term t in T, $\Pi(u) - \{(u)^\bullet\} \subset \Pi(T)$; if x has no occurrence in u, we have also $\Pi(\sigma u) - \{(u)^\bullet\} \subset \sigma^\bullet\Pi(T)$.

If p is in $\Pi(\sigma T)$, there exists a term t in T such that p is in $\Pi(\sigma t)$. The proof is by induction on the structure of t.

a) t is a constant or a variable different from x

$\Pi(\sigma t) = \{(t)\}$; by definition, (t) is in $\Pi(T)$ then $\sigma^\bullet(t) = (t)$ is in $\sigma^\bullet\Pi(T)$ and $\Pi(\sigma t) \subset \sigma^\bullet\Pi(T)$.

b) t=x

** if $\sigma x=u$

$\Pi(\sigma x)=\Pi(u)= (\Pi(u)-\{(u)^\bullet\}) \cup \{(u)^\bullet\}$ and $\Pi(u)-\{u\} \subset \sigma^\bullet\Pi(T)$ since $\sigma u=u$.

Moreover x is a term in T, hence (x) is in $\Pi(T)$ and $\sigma^\bullet(x)$ is in $\sigma^\bullet\Pi(T)$, so that $\Pi(\sigma t) \subset \sigma^\bullet\Pi(T)$.

** if $\sigma x= u+x$

$\Pi(\sigma x)=\Pi(u+x) = (\Pi(u)-\{u\}) \cup \Pi(x) - \{(x)\} \cup \{(u)^\bullet{}_\bullet(x)\}$

We know that $\Pi(u)-\{u\} \subset \sigma^\bullet\Pi(T)$. Obviously $\Pi(x) - \{(x)\}$ is empty. x is a term in T hence (x) is in $\Pi(T)$ and $\sigma^\bullet(x) =(u)^\bullet{}_\bullet(x)$ is in $\sigma^\bullet\Pi(T)$, so that $\Pi(\sigma t) \subset \sigma^\bullet\Pi(T)$.

c) $t= f(t_1,\dots,t_n)$

$\Pi(t)= \{(t)\} \cup (\cup_i \Pi(t_i))$; $\Pi(\sigma t)=\Pi(f(\sigma t_1,\dots,\sigma t_n))= \{(\sigma t)\} \cup (\cup_i \Pi(\sigma t_i))$.

By induction hypothesis on t_i, for every i, $\Pi(\sigma t_i) \subset \sigma^\bullet\Pi(t_i)$. Finally: $\Pi(\sigma t) \subset \{(\sigma t)\} \cup (\cup_i \sigma^\bullet\Pi(t_i)) \subset \sigma^\bullet(\{(t)\} \cup (\cup_i \Pi(t_i))) = \sigma^\bullet\Pi(t) \subset \sigma^\bullet(\Pi(T))$.

d) $t = t_1 + t_2$

$(\sigma t)^\bullet= (\sigma t_1)^\bullet{}_\bullet(\sigma t_2)^\bullet= (s_1,..,s_k)_\bullet(s_{k+1},\dots,s_n)$. We have:

$\Pi(t) =\{(t)^\bullet\} \cup (\Pi(t_1)-\{(t_1)^\bullet\} \cup (\Pi(t_2) -\{(t_2)^\bullet\})$

$\Pi(\sigma t)=\Pi(\sigma t_1+ \sigma t_2) = \{(\sigma t)^\bullet\} \cup (\cup_i(\Pi(s_i)-\{(s_i)\}))$.

By induction hypothesis $\Pi(\sigma t_1) \subset \sigma^\bullet\Pi(t_1)$ and $\Pi(\sigma t_2) \subset \sigma^\bullet\Pi(t_2)$

For every i, $(\Pi(s_i)-\{(s_i)\}$ is included in $\sigma^\bullet\Pi(t_1)-\{(\sigma t_1)^\bullet\}$ or in $\sigma^\bullet\Pi(t_2)-\{(\sigma t_2)^\bullet\}$ and the two prefixes $(\sigma t_1)^\bullet$ and $(\sigma t_2)^\bullet$ are not in $\Pi(s_i)-\{(s_i)\}$.

Finally :

$\Pi(\sigma t) \subset \{(\sigma t)^{\bullet}\} \cup \sigma^{\bullet}(\Pi(t_1)-\{(t_1)\}) \cup \sigma^{\bullet}(\Pi(t_2)-\{t_2^{\bullet}\}) = \sigma^{\bullet}\Pi(t) \subset \sigma^{\bullet}\Pi(T)$. ♦

4. The tree of prefixes

Let T be a finite set of terms , we note $\mathcal{A}(T)$ the following tree:

* The root is ∇ ,
* If $p=(t_1,...,t_n)$ is in $\Pi(T)$, then we have the path $\nabla \to t_1 \to ... \to t_n$ in $\mathcal{A}(T)$.

* If the first i elements of p and p' are the same, we have only one path $\nabla \to t_1 \to ... \to t_n$ in $\mathcal{A}(T)$. If t_{i+1} and t'_{i+1} exist (respectively in p and p') and differ then the node labelled by t_i has two successors (labelled by t_{i+1} and t'_{i+1}).

Remark: Several nodes may have the same label: t may appear in several lists with different prefixes, or t may appear several times in the same list.

Example 3: With $\Pi=\{(a,b,c),(a,b,f(x+y)),(f(t+u)),(a,b,z),(c,z,f(u)),(a,c,a)\}$ we obtain the tree:

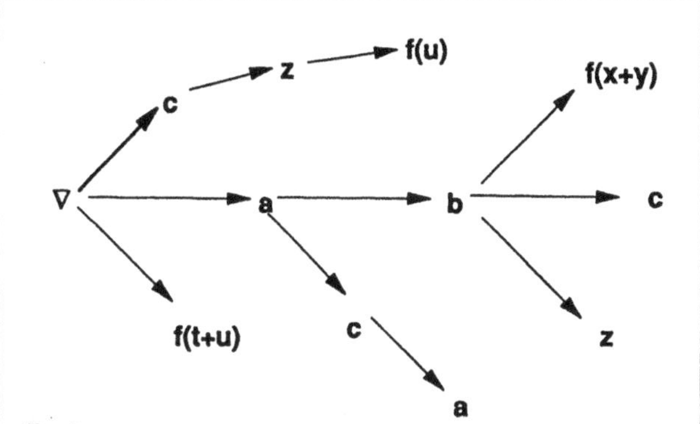

fig 1

5. Unique Prefix Property

5.1. Definition

Definition 5: Given a list $p=(t_1,...,t_{k-1}, x, t_{k+1},....t_n)$, where x is a variable, we say that $(t_1,...,t_{k-1},x)$ is a prefix of x in p. We denote by $\pi(p,x)$ the (possibly empty) set of all the prefixes of x in p.

Definition 6: Given a variable x in Var(T), $\pi(T,x)$ is the set of prefixes of x in $\Pi(T)$ that is the set $\bigcup_{p \in \Pi(T)} \pi(p,x)$.

Definition 7: A set of terms T has the **Unique Prefix Property** (UPP for short) **iff** for each variable x in Var(T), $\pi(T,x)$ is a singleton.

Remark

T has UPP iff every variable in Var(T) appears once and only once in **A(T)**.

5.2. Unification for terms with UPP

Proposition 2: If {t,t'} has UPP then Plotkin's algorithm computes in a finite time a complete set of unifiers of t and t'.

Proof:

We know from Plotkin that his algorithm is complete; hence we have only to deal with the termination.We show that each step of the algorithm corresponds to a particular transformation on the tree of prefixes, so that it becomes evident that UPP is preserved and that only a finite number of such transformations is possible.

Proposition 3: If T= {t,t'} has UPP then so has $\mu \downarrow T$ where μ is the binding computed in one step of the algorithm (if there is no failure !)

Proof:

Every element in $\Pi(T)$ is represented in A(T) by a path starting from the root.

By proposition 1, considering the particular structure of the substitution computed by the algorithm, we know that $\Pi(\mu \downarrow T) = \mu^{\bullet}\Pi(T)$. From this we can deduce the transformations leading from A(T) to $A(\mu \downarrow T)$.

First, we see that if (x, u) is the disagreement pair, there exist in $\Pi(T)$ two prefixes p et p' such that $p=(t_1,\ldots,t_n,x,\ldots.)$ and $p'=(t_1,\ldots,t_n,u_1,\ldots u_k,\ldots)$ where $(u_1,\ldots,u_k)= u^{\bullet}$. and x differs from u (See fig 2 (1)).

We consider two cases:

a) $\sigma = \{x/u\}$

Since x appears only once in the tree, we have only to replace the node labelled by x by a path labelled by $(u_1,...,u_k)$. But there is already such a path in the tree which has the same origin as the edge whose extremity is labelled by x (fig 2 (2)). According to the definition of the tree of prefixes, we must merge them. (fig 2 (3))

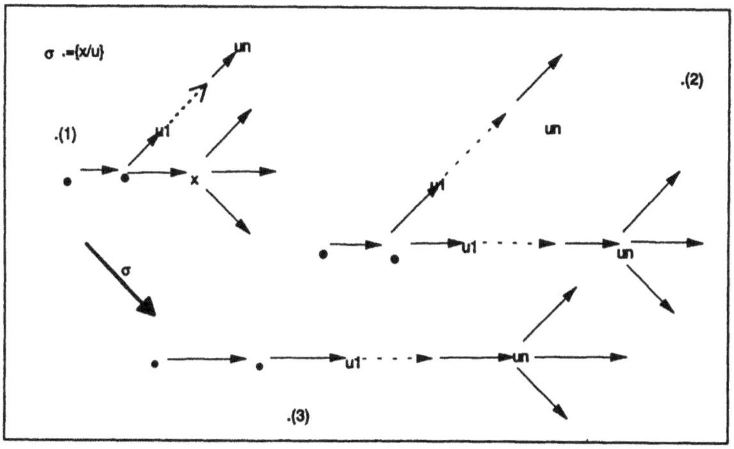

fig 2

The other nodes are neither labelled by x nor by sums. Hence we only have to replace their label l by $\mu\downarrow l$ and possibly merge some of them.

Summing up:

- Some edges and nodes are **moved**
- The node labelled by x is **deleted**.
- Some nodes may be merged.

b) $\sigma = \{x/u+x\}$

We have now to substitute $(u_1,....,u_k,x)$ for x. After the substitution, as in the previous case, we have two paths labelled by $(u_1,...,u_k)$ (fig 3 (4)) with the same origin and we must merge them (fig 3 (5)).

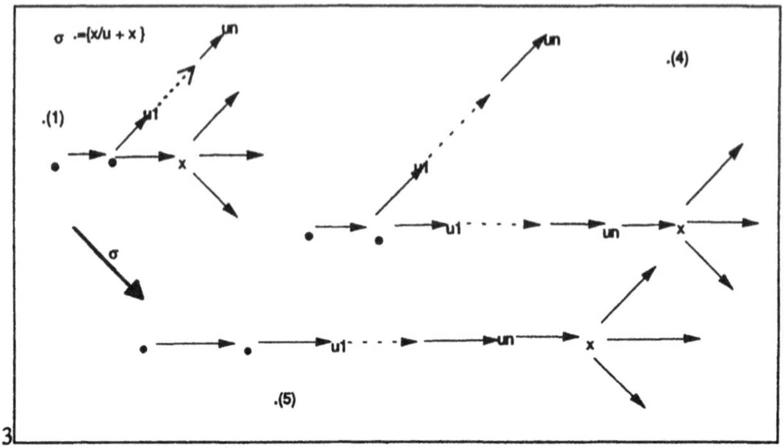

fig. 3

As a result:

- The edge with end labelled by x (and the corresponding subtree) is **moved**, and now hang on to the node labelled by u_n. Remark that the origin of this edge is moved **away** from the root ∇.

- Possibly, some nodes with same prefix and same label are **merged**.

Observe that any variable still occur only once in A(μ↓T) so that μ↓T has UPP. ♦

Now we can complete the proof of Proposition 2.

In the above transformations, the number of edges and nodes of the tree of prefixes **never increase.** Let N be the number of nodes in the initial tree.

- Operation a), which involves (at least) one node deletion, can be performed at most N times.

- Every branch has at most N nodes, so that a node can be moved away from the root at most N-1 times, and operation b) is performed at most N(N-1) times.

It follows that every computation (either succesful or failing) terminates in time $O(N^2)$. In each step, there is a choice between at most three bindings. Hence the tree of all computations of Plotkin's algorithm is finite.

As a consequence, a UPP set of terms has a complete set of unifiers.

Remark: The bound N^2 on the length of a computation can be reached as shown by the following example.

$T=\{f(a+x,a+x,...,a+x), f(x_1+y_1,....,x_n+y_n)\}$. A(t) has 2(n+2) nodes and there is comptutation which requires (n)(n-1)/2 substitutions of the form {y/u+y}. The computed unifier is:

$\{x/a+x_1+x_2+...+x_n+y_n\ ;\ x_i/\ a+x_1+...+x_i\ ;\ ..;\ y_i\ /\ x_{i+1}+...+x_n+y_n\}$.

6. Rules based Unification

It has now become usual to present a unification problem as a set of equations and the unification process as the application of transformations rules, until some "solved form" is obtained. The set of rules and the solved forms depend on the representation of the terms and of course on the equational theory.

In [Debart 92] we extend the previous results to rule-based algorithm using some variants of the rules given in [Jouannaud, Kirchner 90]. We shall not present them here but only pinpoint the problems that arise.

First it is easily seen that we have to make some changes in the definition of the Unique Prefix Property.

For example, it is obvious that the following unification problem P is equivalent to P1 and P2. But the set of terms in P has UPP (as defined previously), and the set of terms in P2 has not!

P={f(a+x,g(a+x+c)=f(a+z+b,g(a +z+c+b)) }   (variables: x,z)

P1={a+x=a+z+b; g(a+x+c)=g(a+z+c+b)}

P2={x=z+t; g(a+x+c)=g(a+z+c+b)}

So we had to modify the definition, and adapt it to this new context. (In fact the new definition coïncides with the previous one for a problem with only one equation).

Secondly, we show that termination depends on the control: a fully non deterministic use of the transformation rules may not terminate, even for a UPP problem (which nevertheless has a finite complete set of unifiers). We establish some rather liberal constraints on the control, sufficient to ensure completeness and termination for UPP problems.

Third, the proposed algorithms consider terms in tree form. For UPP problems the representation of terms and the implementation of the rules may be done using the tree of prefixes. Since every variable appears only once in this tree, we have then the same advantage w.r.t. efficiency as with a DAG representation.

Finally, we extend all the results and algorithms to associative and unitary unification and order-sorted logic. This provides immediately unification algorithms for the theories we used for modal logic.

Bibliography

[Auffray, Enjalbert 89] Y. Auffray P.Enjalbert: *Modal Theorem Proving: An Equational Viewpoint,* Submitted to the Journal of Logic and Computation 1990

[Debart 92] F.Clerin-Debart : *Théories équationnelles et de contraintes pour la déduction automatique en logique modale* Thèse de L'université de Caen 1992

[Debart, Enjalbert,Lescot 90]F.Debart, P.Enjalbert, M.Lescot: *Multi-modal automated Deduction Using Equational And Order-Sorted Logic.* Proc 2nd.Conf on Conditional and typed Rewriting Systems, Montreal 1990. To appear in Springer LNCS, M.Okada&S.Kaplan Ed.1991

[Fages, Huet 86] F.Fages G.Huet *Complete sets of unifiers and matchers in equational theories.* Theor. Computer Science 43, 1986, 189-200.

[Huet 80] G.Huet : *Confluent Reductions. Abstract Properties and Applications to Terms Rewriting Systems.*Journal of the Ass. for Comptuting Machinery. Vol 27 N°4 10-1980

[Jouannaud Kirchner 90] J.P.Jouannaud C.Kirchner : *Solving equations in Abstract Algebras: A rule-based Survey of Unification.* Rapp.de Rech. N°561 CNRS UA 410

[Ohlbach 88] H.J.Ohlbach : *A resolution calculus for Modal Logics* . 9th Int. Conf. on automated deduction. LCNS 310, Springer Verlag 1988

[Plotkin 72] G.D.Plotkin : *Building-in Equational Theory.*Machine Intelligence 7, 73-90 ,1972

[Plotkin] G.D.Plotkin : *Some Unification Algorithms.* Research Memorandum. School of Artificial Intelligence, University of Edinburgh

Theorem proving by combinatorial optimization

Hachemi BENNACEUR and Gérard PLATEAU
Université Paris-Nord
Institut Galilée
Laboratoire d'Informatique de Paris-Nord
avenue J.B.Clément - 93430 Villetaneuse - France
tel: 1/49403573 fax:1/49403333
email:gp@lipn.univ-paris13.fr

Abstract:

The inference problem in propositional logic realizes a strong connection between Artificial Intelligence and Operational Research. It is now well-known that this problem can be formulated as a constraint satisfaction problem (CSP), whose system has a generalized covering type (we recall that a CSP consists in proving the emptiness of a domain defined by a set of diophantine constraints, or the existence of a solution). We propose a new method -denoted by FAST (Fast Algorithm for the constraint Satisfaction Test problems)- which allows an efficient solution of the CSP instances for logical inference. Computational results are reported for 3-SAT, 4-SAT and system expert type instances.

Key Words : Constraint Satisfaction Problem - Logical Inference -Integer Programming - Branch and Cut - Automatic Theorem Proving - Combinatorial Problem - Design of Algorithm

1 Introduction

The inference problem in propositional logic can be modelized as a system of diophantine constraints for which the existence of a solution or the emptiness of the associated domain has to be proved.

Although this constraint satisfaction problem -denoted by CSP- is NP-complete, it is possible to design efficient exact methods for several class of such instances. Blair, Jeroslow and Lowe [4] and more recently Hooker [7,8,9] have obtained good results using integer programming tools. Their computational experiments show clearly that their algorithms are largely faster than the classical previous resolution algorithms for satisfiability and inference problems.

Hooker's method consists in adding an objective function to the constraint system in order to solve an equivalent O-1 programming problem by a branch and cut method. We propose a new method -denoted by FAST- which allows an efficient solution of the CSP instances for logical inference problem. The main characteristics of our method is the solving of a sequence (very short in practice) of 0-1 programming problems. Each generic problem of this sequence has an appropriate objective function and a constraint system size lower than the initial CSP's one (very much lower in practice).

This paper is organized as follows:

Section 2 gives the modelization of inference problems in terms of CSP whose system has a generalized covering type. Section 3 deals with our exact algorithm FAST for the CSP [1,2,3] . Section 4 details the computational results on a SUN 3/160 computer with a Fortran-77 implementation of FAST performed with a lot of randomly generated logical inference instances.

2 Modelization

Given a logical formula in conjunctive normal form

$$F = \wedge_{i=1}^{m} C_i$$

where each clause C_i is the disjunction of a subset of n given literals t_j:

$$C_i=\vee_{j\in J_i} t_j \ , \ J_i \subseteq \{1,2,\ldots,n\}, \quad i=1,\ldots,m,$$

the equivalence between the satisfiability (SAT) problem associated with F and the constraint satisfaction problem is shown below:

The satisfiability of F consists in solving the boolean equation

$$F=\wedge_{i=1}^{m} C_i \text{ is true}$$

which means

$$C_i \text{ is true} \quad i=1,\ldots,m, \text{ that is } \vee_{j\in J_i} t_j \text{ is true} \quad i=1,\ldots,m.$$

Let us denote by x_j, $j=1,\ldots,n$, binary variables associated with the literals t_j in the following manner:

$x_j=1$ if t_j is true in F, and $x_j=0$ otherwise.

Thus, C_i is true $\Leftrightarrow \sum_{j\in J_i} y_j \geq 1 \quad i=1,\ldots,m$

with $y_j = x_j$ if t_j is a positive literal,
$ = 1-x_j$ otherwise.

By denoting $J_i^+ = \{ j\in J_i \mid t_j \text{ is a positive literal}\}$ and $J_i^- = \{ j\in J_i \mid t_j \text{ is a negative literal}\}$,

we can conclude that:

$$C_i \text{ is true} \Leftrightarrow \sum_{j\in J_i^+} x_j + \sum_{j\in J_i^-} (1-x_j) \geq 1, \text{ that is } \sum_{j\in J_i^+} x_j - \sum_{j\in J_i^-} x_j \geq 1 - |J_i^-|$$

or

$$\sum_{j\in J_i^+} x_j - \sum_{j\in J_i^-} x_j \geq b_i \quad \text{with} \quad b_i = 1 - |J_i^-| \quad i=1,\ldots,m.$$

By defining a matrix $A=(a_{ij})$ such that:

$a_{ij}=1$, if $j\in J_i^+$; $a_{ij}=-1$, if $j\in J_i^-$; $a_{ij}=0$, if $j \notin J_i$ $\quad i=1,\ldots,m$,

and by noting that all the components of the right-hand side b of the constraints are integer numbers, and each coefficient of the left-hand side A of the constraints belongs to $\{-1, 0, 1\}$, we deduce the

Property 1.

The Satisfiability Problem is solved by proving the feasability of a system

$$Ax \geq b; \; x \text{ binary}$$

corresponding to a generalized covering problem, or the emptiness of the associated domain.

Illustration:

Expert system is a typical illustration : the rules and the facts are defined as logical inferences which can always be reformulated as a set of clauses.

Example: (M. Gondran [6])

Given the following knowledge base

base of facts: $\{H, K\}$

base of rules:

(R1) $A \Rightarrow E$

(R2) $B \Rightarrow D$

(R3) $H \Rightarrow A$

(R4) $E \wedge G \Rightarrow C$

(R5) $E \wedge K \Rightarrow B$

(R6) $D \wedge E \wedge K \Rightarrow C$

(R7) $G \wedge K \wedge F \Rightarrow A$

where "$A \Rightarrow E$" means " from A we can deduce E ",

theorem proving consists in the induction of news facts from the knowledge base using the modus ponens rule ("if p is true and $p \Rightarrow q$ then q is true").

Let us assume that we want to prove $E \wedge F$. This goal is reached by proving the truth of the logical formula

$$H \wedge K \wedge R1 \wedge R2 \wedge R3 \wedge R4 \wedge R5 \wedge R6 \wedge R7 \Rightarrow E \wedge F .$$

This is equivalent to solve a satisfiability problem associated with a logical formula whose clauses are distributed in the three subsets:

clauses associated with the facts: $\{ H; K \}$

clauses associated with the rules: $\{ \neg A \vee E; \neg B \vee D; \neg H \vee A;$

$\neg E \vee \neg G \vee C; \neg E \vee \neg K \vee B; \neg D \vee \neg E \vee \neg K \vee C; \neg G \vee \neg K \vee \neg F \vee A \}$

clause which defines the goal: $\{\neg E \vee \neg F\}$.

The theorem is valid if this SAT problem is unsatisfiable, thus if the following equivalent 0-1 programming model is infeasible:

by assigning the binary variables $x_1, x_2, \ldots, x_9$ to the nine literals in the alphabetical order A,B,...,K, we deduce the 0-1 linear system

$x_8=1;\ x_9=1$ {base of facts}
$1-x_1+x_5 \geq 1$ {base of rules}
$1-x_2+x_4 \geq 1$
$1-x_8+x_1 \geq 1$
$1-x_5+1-x_7+x_3 \geq 1$
$1-x_5+1-x_9+x_2 \geq 1$
$1-x_4+1-x_5+1-x_9+x_3 \geq 1$
$1-x_7+1-x_9+1-x_6+x_1 \geq 1$
$1-x_5+1-x_6 \geq 1$ {goal}
$x_j \in \{0,1\},\ j=1,\ldots,9$

which has at least the solution $x_1=1$, $x_2=1$, $x_3=1$, $x_4=1$, $x_5=1$, $x_6=0$, $x_7=0$, $x_8=1$, $x_9=1$. Thus the theorem is not valid. This means that the goal $E \wedge F$ cannot be deduced from the knowledge base.

3 Description of the method FAST

This part deals with the description of our exact algorithm - denoted by FAST - for the constraint satisfaction problem. The principle of algorithm FAST (section 3.1) and a numerical illustration (section 3.2) precede the basic theoretical results (section 3.3) and the description of our algorithm (section 3.4).

3.1 Principle of the method

The method FAST is designed for the exact solving of the constraint satisfaction problem (see [1,2,3] for more details and other applications). Its aim is the proof of the emptiness of a domain defined by a system of m linear diophantine constraints, or the existence of a solution; namely, this system is supposed to be written in the canonical form:

$$(S) \qquad Ax \le b;\ x \in \mathbf{D}$$

where A is a mxn matrix, b is an integer vector, and **D** is a finite domain of integer points.

Given an initial integer point x^0, the scheme of FAST consists in generating a finite sequence of integer points

$$x^1, x^2, \ldots, x^k \qquad \text{where } k \le m$$

until, either x^k satisfies the system of constraints (S), or the associated domain ($F(S)=\{x \in \mathbf{D} \mid Ax \le b\}$) is proved to be empty.

Each integer point x^h ($h \in \{1,\ldots,k\}$) is an optimal and feasible solution (or the current best feasible solution for x^k) of an integer programming problem whose constraints are the ones satisfied by x^{h-1}, and objective function is the sum of the current unsatisfied constraints. In addition, the aim of the optimization is to go inside the domain F(S), namely it is such that the set of satisfied constraints grows strictly from one iteration to the next one.

3.2 Illustration

At each iteration, by denoting L the subset of the m constraints of (S) already satisfied, and G the set of the other constraints, the current generic problem has the following form:

$$(P(L,G)) \qquad \begin{array}{ll} \min \sum_{i \in G} A_i x & \\ \text{s.t.} \quad A_i x \le b_i & i \in L \\ \quad x \in \mathbf{D} & \end{array}$$

Letting x^* be an optimal solution of (P(L,G)), and v its value, the following example illustrates the general principle of our method FAST

$$(S) \qquad \begin{array}{lll} 2x_1 + 3x_2 & \le 6 & (1) \\ 2x_1 - 3x_2 & \le 3 & (2) \\ -3x_1 - x_2 & \le -5 & (3) \\ \multicolumn{3}{l}{x \in \mathbf{D} = \{ x \in \mathbf{Z}^2 \mid x_1 \ge -1;\ x_2 \le 3 \}} \end{array}$$

<u>initialization</u>

By letting $L = \emptyset$ and $G=\{1,2,\ldots,m\}$, the starting point x^0 may be choosen by solving

$$(P(L,G)) \qquad \min \; x_1-x_2 \quad \text{s.t.} \quad x \in \mathbf{D}$$

whose value and solution are $v=-4$ and $x_1^0=-1$, $x_2^0=3$,

thus $L=\{2\}$ and $G=\{1,3\}$.

Note: as the right hand-side of constraints (1) and (2) are non negative, the initial point x^0 might have been the origin. But we have taken the above choice for the sake of explanation in order to show several iterations.

<u>Iteration 1</u>

$$(P(L,G)) \qquad \begin{array}{ll} \min & -x_1+2x_2 \\ \text{s.t.} & 2x_1-3x_2 \leq 3 \\ & x \in \mathbf{D} \end{array}$$

the value and solution of (P(L,G)) are $v=-1$ and $x_1^1=-1$, $x_2^1=-1$,

thus $L=\{1,2\}$ and $G=\{3\}$.

<u>iteration 2</u>

$$(P(L,G)) \qquad \begin{array}{ll} \min & -3x_1-x_2 \\ \text{s.t.} & 2x_1+3x_2 \leq 6 \\ & 2x_1-3x_2 \leq 3 \\ & x \in \mathbf{D} \end{array}$$

the value and solution of (P(L,G)) are $v=-4$ and $x_1^2=1$, $x_2^2=1$.

As the minimal value of the left-hand side of constraint (3) subject to the constraints (1) and (2), is strictly greater than its right-hand side, we can conclude that F(S) is empty.

3.3 Basic theoretical results

Given the generic problem (P(L,G)) defined in section 3.2 (F(P(L,G)) denotes its domain, x^* an optimal solution, v its value, and OS(P(L,G)) its set of optimal solutions), the following results give stopping criteria for the method FAST:

Theorem 1.
If one of the following conditions holds

(i) $v > \sum_{i \in G} b_i$

(ii) $v = \sum_{i \in G} b_i$ *, and*

$\forall\, x^*$ *optimal solution of* $(P(L,G))$ $: \exists\, i \in G$ *such that* $A_i x^* \neq b_i$

then $F(S)$ *is empty.*

Theorem 2.
If $A_i x^* \leq b_i \quad \forall\, i \in G$ *then* $x^* \in F(S)$.

Corollary 1.
If the conditions of theorems 1 and 2 do not hold then:

(i) $v < \sum_{i \in G} b_i$

(ii) $\exists\, i \in G : A_i x^* > b_i$ *and* $\exists\, j \in G : A_j x^* < b_j$.

If the conditions of theorems 1 and 2 do not hold, corollary 1 (ii) implies that the cardinality of the set $I = \{\, i \in G \mid A_i\, x^* \leq b_i \,\}$ is at least equal to 1, and the following results are used:

Theorem 3.
If the conditions of theorems 1 and 2 do not hold for (P(L,G)), let y^* *be the optimal solution of the next generic problem* $(P(L \cup I, G \setminus I))$ *in the sequence, then*

$$F(S) \neq \emptyset \Rightarrow \sum_{i \in I} A_i y^* > \sum_{i \in I} A_i x^* . \qquad (1)$$

Corollary 2.
If the conditions of theorems 1 and 2 do not hold and $\forall i \in I \;\; A_i x^* = b_i$ *then* $F(S) = \emptyset$.

3.4 Algorithm FAST

The algorithm FAST for the exact solving of the constraint satisfaction problem is based on the theoretical results described in section 3.3.

```
{goal: to prove the existence of a solution for a given system (S) Ax ≤ b; x∈ D , or
the emptiness of its domain F(S)}
choose x0 in D;

if x0 satisfies (S) then x0 is a solution of (S)
else
    L← set of constraints of (S) satisfied by x0;
    G← set of the other constraints of (S);
    x* ← optimal solution of the generic problem (P(L,G));
    while goal is not reached do
        if condition (i) of theorem 1 holds then the domain is empty
        else I ← set of the new constraints satisfied by x*
            if I= G then x* is a solution of (S)
            else if condition (ii) of theorem 1 holds
                then the domain is empty
                else L←L ∪ I ; G← G\I ;
                    y* ← optimal solution of the generic problem (P(L,G))
                    if condition of theorem 3 holds
                    then      the domain is empty
                    else      x* ← y*
                    endif
                endif
            endif
        endif
    endwhile
endif
```

Proposition.
The number of iterations of algorithm FAST is bounded by m-1.
Proof.
Direct consequence of corollary 1 which proves that at each iteration $|I| \geq 1$ and thus $|G|$ decreases by at least one unity. In the worst case the algorithm stops with the

solving of a problem whose objective function is reduced to one constraint of (S) subject to the m-1 other constraints of (S). ♦

3.5 Application to logical inference:

At each iteration of method FAST, the current generic problem (P(L,G)):

$$\min \; cx$$

$$\text{s.t.} \quad \sum_{j\in J_i^-} x_j - \sum_{j\in J_i^+} x_j \leq |J_i^-| - 1 \quad i \in L$$

$$x_j \in \{0,1\},\; j=1,\ldots,n,$$

where c is the sum of the left-hand side of the unsatisfied constraints, is solved by a branch and cut method. At each node of the enumeration tree, we apply several tests allowing to reduce the size (fixation of variables and elimination of redundant constraints) of the associated subproblem. The structure of the model is exploited in order to design heuristics and cuts; duality in integer programming [11] allows to construct good relaxations. As we use a branch and cut method it is possible to find a solution of (S) before the complete solving of the generic problem (P(L,G)). In this case, it is an obvious additional stopping criterion for our method FAST.
In addition, the following result directly deduced from the structure of the generalized covering system, is important for the resolution of the CSP:

Property 2.

A constraint $\sum_{j\in J_i^-} x_j - \sum_{j\in J_i^+} x_j \leq |J_i^-| - 1 \quad i\in\{i,\ldots,m\}$ *is satisfied for all binary vector x such that:* $\exists j\in J_i^+ : x_j=1$ *or* $\exists j\in J_i^- : x_j=0$.

It allows the quick detection of the feasibility of a given solution, and it allows to point out easily the redundant constraints after fixing a subset of variables.

4 Computational experiments

The Fortran-77 code of algorithm FAST has been implemented on a SUN 3/160 computer with a lot of instances of the generalized covering type

$Ax \geq b;\; x \in \{0,1\}^n$

with $A \in \{0,1,-1\}^{m\times n}$ and $b \in \mathbf{Z}^m$

randomly generated with the Purdom and Brown model [13]. Each clause is randomly and independently generated with a fixed number k of literals. (Each literal has the same probability; each clause includes distinct literals).
Three classes of instances are considered: the classes 1 and 2 are associated respectively with k=3 and k=4 (satisfiability problems) (tables 1 and 2); the class 3 concerns typical instances deduced from expert systems (logical inference)(table 3). For each type of instances (e.g. given the number of clauses m and the number of variables n), the tables detail the average CPU times in seconds over ten problems.
The table 3 concerns 50-variables instances and points out the number of unary clauses (e.g. base of facts (bf)), binary clauses (e.g. base of rules with 2 literals (br2)), ternary clauses (e.g. base of rules with 3 literals (br3)), and finally clauses with 4 literals (br4).

m	n	time
250	20	1.34
250	30	4.5
250	40	17.4
250	50	77
250	60	137
250	80	64
250	100	38
250	150	8

table1: satisfiability problems with k=3

m	n	time
150	30	1.58
200	30	1.19
250	30	17.94
300	30	24
350	30	40.7
400	30	34
600	30	28.96
700	30	25.26
800	30	26.79
1000	30	27.91

table 2: satisfiability problems with k=4

m	bf	br2	br3	br4	time
100	20	40	20	20	1.2
200	40	60	50	50	2.2
400	100	50	150	100	4.13
600	100	50	200	250	6.17
800	20	80	200	500	8.19

table 3: inference problems

5 Conclusion

Although the CSP is a NP-complete problem, the computational experiments detailed in section 4 show the efficiency of the method FAST. The computational results obtained by Blair, Jeroslow and Lowe [4] , and more recently by Hooker [7,8,9], show that integer programming resolution is more robust and efficient than the symbolic resolution method (such as set-of-support resolution [12]). The dominance is often very large: in [9], there is a lot of instances for which CPU time is a few seconds for integer programming, and more than ninety minutes for symbolic resolution. Our method which exploits adequate integer programming tools, confirms these results. The efficiency of our exact method is explained by the very small number of iterations, although it uses for each generic problem a branch and cut method. Other experiments are in progress to complete the study of the effective impact of using FAST in the satisfiability and inference problems. In addition the incremental aspect of our method allows its use in the context of constraint logic programming.

References

[1] H. Bennaceur, Le problème de satisfaction de contraintes synthèse et méthode exacte de résolution, Thèse de Doctorat, Université Paris-Nord, France, (1989)

[2] H. Bennaceur and G. Plateau, Sur le problème de satisfaction de contraintes, Research report LIPN 89-6, Université Paris-Nord, France (1989)

[3] H. Bennaceur and G. Plateau, Le problème de satisfaction de contraintes: modèles des applications informatiques et méthode de résolution exacte, ResearchReport LIPN 90-4, Université Paris-Nord, France (1990)

[4] C. E. Blair, R. G. Jeroslow and J. K. Lowe, Some results and experiments in programming techniques for propositional logic, Computers and Operations Research (1988)

[5] M. Davis and H. Putnam, A computing procedure for quantification theory, Journal of the ACM 7 (1960) 201-215

[6] M. Gondran, Intoduction aux systèmes experts, Ed. Eyrolles (1984)

[7] J. N. Hooker and C. Fedjki, Branch-and-cut solution of inference problems in propositional logic, Research report Carnegie-Mellon University Pittsurgh, Pennsylvania 15213 (1987)

[8] J. N. Hooker, A quantitative approach to logical inference, Decision Support Systems 4 (1988) 45-69.

[9] J. N. Hooker, Resolution vs. cutting plane solution of inference problems: some computational experience, Research report Carnegie-Mellon University Pittsurgh, Pennsylvania 15213 (1988)

[10] R. G. Jeroslow and J. Wang, Solving propositional satisfiability problems, Research report Georgia Institute of Technology, Atlanta, GA (1987)

[11] G. L. Nemhauser and L. A. Wolsey, Integer and combinatorial optimisation, Willey-Interscience (1988)

[12] N.J. Nilsson, Principles of Artificial Intelligence, Tiogo, Palo Alto, CA, 1980.

[13] P. W. Purdom and C. A. Brown, Polynomial average-time satisfiability problems, India University Computer Science, Technical Report 118 (1981)

Solving String Equations with Constant Restrictions

Peter Auer

Institut für Computergraphik, Technical University Vienna, A–1040, Austria

January 24, 1992

Abstract

We present a variant and extension of Makanin's [5] decision algorithm for string equations: Given are sets $\mathcal{C} = \{c_1, \ldots, c_m\}$ of constants, $V = \{x_1, \ldots, x_n\}$ of variables and a string equation $s_1 \asymp s_2$ where $s_1, s_2 \in (\mathcal{C} \cup V)^+$. Furthermore sets $R(x_i) \subseteq \mathcal{C}$, $1 \leq i \leq n$, are given which are called constant restrictions. A substitution σ solves the equation $s_1 \asymp s_2$ and satisfies the constant restrictions $R(x_i)$, $1 \leq i \leq n$, if $\sigma(s_1) = \sigma(s_2)$ and $\sigma(x_i) \in ((\mathcal{C} - R(x_i)) \cup V)^+$ for all $x_i \in V$. I.e. we consider solutions of string equations such that certain constants do not appear in the substitutions of some variables.

Modifying the decision algorithm of Makanin we obtain an algorithm which decides whether or not a given string equation has a solution satisfying the constant restrictions. Furthermore we think that we have, as a by-product, a very nice presentation of Makanin's algorithm.

1 Introduction

In this paper we consider string equations with constant restrictions which are defined as follows.
Let $\mathcal{C}$ be a finite set of constants and $\mathcal{V}$ a set of variables, $\mathcal{C} \cap \mathcal{V} = \emptyset$. The set of strings is given as $\mathcal{S} = (\mathcal{C} \cup \mathcal{V})^+$ ([1]) and a string equation is denoted by $s_1 \asymp s_2$ where $s_1, s_2 \in \mathcal{S}$. A substitution is a function $\sigma : \mathcal{S} \to \mathcal{S}$ with

$$\begin{aligned} \sigma(a) &= a \quad \forall a \in \mathcal{C} \\ \sigma(st) &= \sigma(s)\sigma(t) \quad \forall s, t \in \mathcal{S}. \end{aligned}$$

Clearly σ is determined by all $\sigma(x)$, $x \in \mathcal{V}$. A substitution solves a string equation $s_1 \asymp s_2$ if $\sigma(s_1) = \sigma(s_2)$.
The problem of deciding if there is a substitution solving a given string equation was open for a long time and was finally solved by Makanin [5], who was able to construct a decision algorithm. For variants of his algorithm see [4], [7], [8].
Clearly the problem of solving string equations is equivalent to the unification problem in associative theories $E = \{f(x, f(y, z)) \equiv f(f(x, y), z)\}$ where terms t_1, t_2 are unifiable

[1] Formally $\mathcal{S} = \bigcup_{n \geq 1} \mathcal{S}_n$ where $\mathcal{S}_1 = \mathcal{C} \cup \mathcal{V}$ and $\mathcal{S}_{n+1} = \{st : s \in \mathcal{S}_n, t \in \mathcal{S}_1\}$.

modulo E if there is a substitution σ with $\sigma(t_1) \equiv_E \sigma(t_2)$. Since during the last years the combination of unification algorithms received a lot of interest we wanted to combine Makanin's algorithm with unification algorithms for other equational theories. By a result of Baader, Schulz [3] and Auer [1] this is possible if we find an algorithm which accepts as input

1. a finite set of string equations P
2. for all variables x in P a set $R(x) \subseteq \mathcal{C}$

and outputs

- YES if there is a substitution σ which solves all string equations in P and satisfies $\sigma(x) \in ((\mathcal{C} - R(x)) \cup \mathcal{V})^+$
- NO otherwise.

The set $R(x)$ is called constant restriction for variable x. A substitution satisfies the constant restriction $R(x)$ if $\sigma(x) \in ((\mathcal{C} - R(x)) \cup \mathcal{V})^+$, i.e. $\sigma(x)$ contains no constant from $R(x)$.
The sets of variables appearing in a string s, string equation $s_1 \asymp s_2$, set of string equations P will be denoted by var(s), var$(s_1 \asymp s_2)$, var(P), respectively. Analogously the sets of constants will be denoted by const(s), const$(s_1 \asymp s_2)$, const(P).
In the following sections we will present and modify Makanin's decision algorithm such that it satisfies the above conditions, so that we have the following

Theorem 1 *There is an algorithm deciding if for a given set of string equations P and given constant restrictions $R(x) \subseteq \mathcal{C}$, $x \in$ var(P), there is substitution σ which solves all equations in P and satisfies the constant restrictions $R(x)$.*

Remark 1 *We learned recently that a similar and even more general extension of Makanin's algorithm was given by Schulz [8]*

2 Generalized equations

Consider a pair (P, R) where P is a finite set of string equations and $R = \{R(x) : x \in$ var$(P)\}$ is a set of constant restrictions. In this section we transform the pair (P, R) into a generalized equation which is a special structure to decide string equations. It was invented by Makanin [5], [6]. We will work with a modification of it, using improvements made by Jaffar [4] and adding the constant restrictions in some way.

Definition 1 (Generalized equations) *A generalized equation is a tupel*

$$(BOUND, VAR, BASE, CONN, REST, REP)$$

where

BOUND is a finite set of boundaries $B \subseteq \mathcal{V}$, *which are partially ordered by a relation* $<$ *with a minimum element* $b_0 \in B$ *such that* $b_0 \leq b$ *for all* $b \in B$.

VAR is a finite set of variable symbols $V \subseteq \mathcal{V}$, $V \cap B = \emptyset$, *with a self-inverse function* $\Delta : V \to V$ *such that*

$$\Delta x \neq x, \Delta\Delta x = x \quad \forall x \in V.$$

BASE is a finite set of base equations *where a base equation is of the form* $b_l x \asymp b_r$, $b_l, b_r \in B$, $x \in V \cup C \cup \{\epsilon\}$. *If* $x = \epsilon$ *then* $b_l = b_r$, *if* $x \neq \epsilon$ *then* $b_l < b_r$. *Each variable* $x \in V$ *must appear exactly once in BASE.*

CONN is a finite set of boundary connections *where a boundary connection is a triple* $(b_1/x/b_2)$, $b_1, b_2 \in B$, $x \in V$ *with* $b_l(x) < b_1 < b_r(x)$, $b_l(\Delta x) < b_2 < b_r(\Delta x)$ *where* $b_l(x)x \asymp b_r(x), b_l(\Delta x)\Delta x \asymp b_r(\Delta x) \in BASE$.

REST is a finite set of modified restrictions *where a modified restriction is a triple* $(b_1, \neg a, b_2)$, $b_1, b_2 \in B$, $b_1 < b_2$, $a \in C$ *such that a appears in BASE.*

$REP \geq 1$ *is a natural number.*

Definition 2 (Left and right boundaries) *All boundaries* b_l *which appear in a base equation* $b_l x \asymp b_r \in BASE$ *are called left boundaries, all boundaries* b_r *are called right boundaries.*
Since each $x \in V$ *appears exactly once in BASE we can define unique functions* $\beta_l, \beta_r : V \to B$ *such that* $\beta_l(x)x \asymp \beta_r(x) \in BASE$ *for all* $x \in V$.

Definition 3 (Solution of a generalized equation) *A substitution* σ *is a solution of a generalized equation* $(BOUND, VAR, BASE, CONN, REST, REP)$ *if*

C_{BOUND}: $b_1 < b_2$ *implies* $\sigma(b_1)s = \sigma(b_2)$ *for some* $s \in \mathcal{S}$.

C_{VAR}: $\sigma(x) = \sigma(\Delta x)$ *for all* $x \in V$.

C_{BASE}: σ *solves all base equations in BASE.*

C_{CONN}: *for all* $(b_1/x/b_2) \in CONN$ *there exists a string* $s \in \mathcal{S}$ *such that*

$$\sigma(\beta_l(x))s = \sigma(b_1), \quad \sigma(\beta_l(\Delta x))s = \sigma(b_2),$$

C_{REST}: *for all* $(b_1, \neg a, b_2) \in REST$ *there is a string* $s \in \mathcal{S}$ *such that* $\sigma(b_1)s = \sigma(b_2)$ *and* $s \in ((C - \{a\}) \cup \mathcal{V})^+$.

Remark 2 *Since the boundary connection* $(b_1/x/b_2)$ *means that a solution* σ *must satisfy*

$$\sigma(\beta_l(x))s = \sigma(b_1), \quad \sigma(\beta_l(\Delta x))s = \sigma(b_2)$$

and $(b_2/\Delta x/b_1)$ *implies*

$$\sigma(\beta_l(\Delta x))s = \sigma(b_2), \quad \sigma(\beta_l(x))s = \sigma(b_1)$$

these boundary connections are equivalent. Therefore we will treat them as one unique boundary connection, i.e. we set $(b_1/x/b_2) = (b_2/\Delta x/b_1)$ *for all boundary connections* $(b_1/x/b_2)$.

Definition 4 *The substring relation* $\ll$ *on strings is defined as*

$$\begin{aligned} s \ll t \iff & t = s \quad or \\ & t = su \quad or \\ & t = us \quad or \\ & t = usv \end{aligned}$$

where $s, t, u, v \in \mathcal{S}$.

The length of a string $s \in \mathcal{S}$ *denoted by* $|s|$ *is defined as*

$$\begin{aligned} &|x| = 1 \quad \forall x \in \mathcal{C} \cup \mathcal{V} \\ &|s_1 s_2| = |s_1| + |s_2|. \end{aligned}$$

Definition 5 (Repetitions) *Let* ϵ *be the empty string such that* $\epsilon s = s\epsilon = s$ *for all* $s \in \mathcal{S}$. *If* $s \in \mathcal{S}$ *then the n-fold repetition* s^n *is defined as*

$$\begin{aligned} &s^0 = \epsilon, \\ &s^n = ss^{n-1} \quad \forall n \geq 1. \end{aligned}$$

The predicate $\mathrm{R}\#(\sigma, V, r)$ *where* σ *is a substitution,* $V \subseteq \mathcal{V}$, $r \geq 1$, *is defined as follows:*

$$\mathrm{R}\#(\sigma, V, r) \iff \forall x \in V \quad \nexists s \in \mathcal{S} : s^{r+1} \ll \sigma(x).$$

Remark 3 $\mathrm{R}\#(\sigma, V, r)$ *means that* $\sigma(x)$, $x \in V$, *contains only an at most r-fold repetition of a substring.*

Example 1. Let $\sigma(x) = abcbcbca$. Then $\mathrm{R}\#(\sigma, \{x\}, r)$ holds for $r \geq 3$ but does not hold for $r < 3$.

□

Definition 6 (Short solution) *A solution* σ *of a generalized equation*

$$(BOUND, VAR, BASE, CONN, REST, REP)$$

is called short *if* $\mathrm{R}\#(\sigma, V, REP)$ *holds.*

Since we need the proof of the intended algorithm to give a full motivation of the above definitions, we only illustrate the meaning of a generalized equation by an example.

2.1 Relation between string equations and generalized equations

Consider the string equation

$$xcxcx \asymp ayyb$$

with the restriction $R(x) = \{c\}$, where $a, b, c \in \mathcal{C}$ and $x, y \in \mathcal{V}$. This equation can be drawn as a picture:

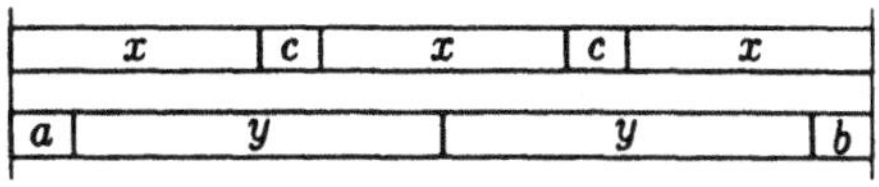

If we mark the begin and the end of the symbols with boundary symbols b_i

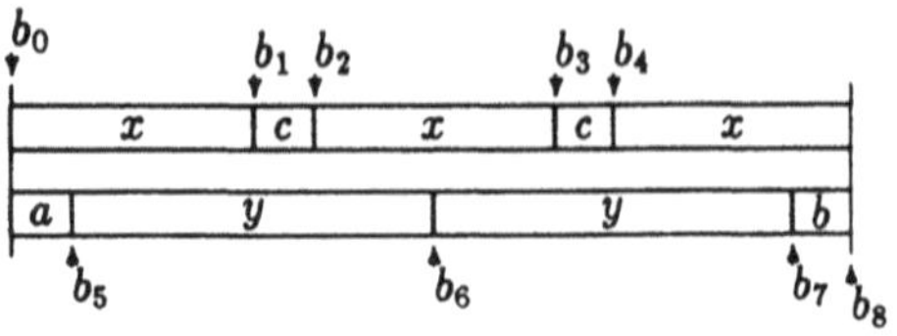

we get the set of boundaries $B = \{b_0, \ldots, b_8\}$ with the partial order $b_0 < b_1 < b_2 < b_3 < b_4 < b_8$ and $b_0 < b_5 < b_6 < b_7 < b_8$. The base equations are given by

$$\begin{array}{lllll} b_0x \asymp b_1 & b_1c \asymp b_2 & b_2x \asymp b_3 & b_3c \asymp b_4 & b_4x \asymp b_8 \\ b_0a \asymp b_5 & b_5y \asymp b_6 & b_6y \asymp b_7 & b_7b \asymp b_8. & \end{array}$$

If a substitution σ solves these equations then clearly

$$\begin{aligned} \sigma(b_0xcxcx) &= \sigma(b_1cxcx) \\ &= \sigma(b_2xcx) \\ &\ \ \ldots \\ &= \sigma(b_8) \\ &= \sigma(b_7b) \\ &\ \ \ldots \\ &= \sigma(b_0ayyb) \end{aligned}$$

and hence $\sigma(xcxcx) = \sigma(ayyb)$.
Since x and y may appear only once in $BASE$ we have to introduce new variables x_1, x_2, x_3, y_1, y_2:

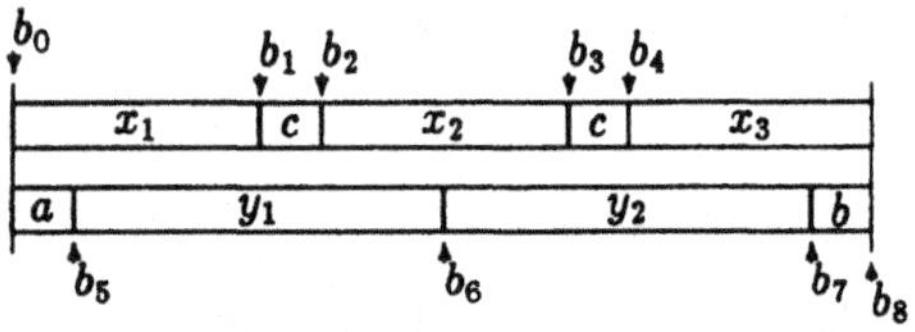

We must guarantee that a solution σ satisfies $\sigma(x_1) = \sigma(x_2) = \sigma(x_3)$ and $\sigma(y_1) = \sigma(y_2)$. This can be done by the function Δ if we define

$$\Delta x_1 = x_2, \ \Delta y_1 = y_2.$$

To guarantee that $\sigma(x_1) = \sigma(x_3)$ we need an additional variable x_4. We define

$$\Delta x_3 = x_4$$

and add x_4 to the picture:

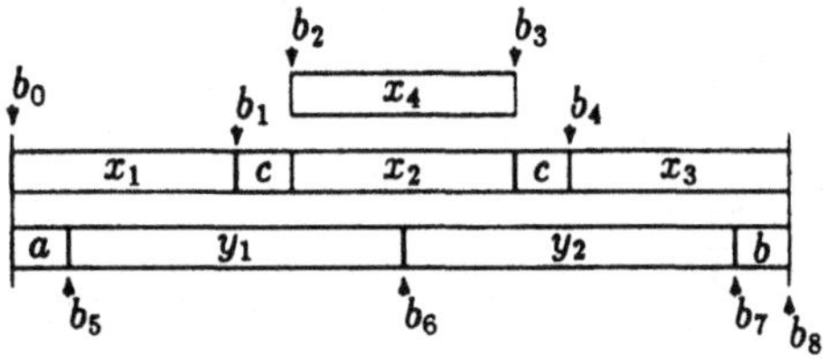

Then we have

$$BASE = \{ \quad b_2x_4 \asymp b_3, \\ b_0x_1 \asymp b_1, b_1c \asymp b_2, b_2x_2 \asymp b_3, b_3c \asymp b_4, b_4x_3 \asymp b_8, \\ b_0a \asymp b_5, b_5y_1 \asymp b_6, b_6y_2 \asymp b_7, b_7b \asymp b_8\}.$$

If σ is a solution then $\sigma(b_2x_2) = \sigma(b_3) = \sigma(b_2x_4)$ implies $\sigma(x_1) = \sigma(\Delta x_1) = \sigma(x_2) = \sigma(x_4) = \sigma(\Delta x_3) = \sigma(x_3)$. Hence the generalized equation has a solution if and only if the equation $xcxcx \asymp ayyb$ has a solution.
The restriction $R(x) = \{c\}$ is transformed into the modified restriction $(b_0, \neg c, b_1)$ which implies $c \not\leq \sigma(x_1)$ since $\sigma(b_0)\sigma(x_1) = \sigma(b_1)$. The set of boundary connections is empty.
To decide whether a pair (P, R) has a solution or not, we will transform it into an equivalent generalized equation. Then this generalized equation will be manipulated until we can trivially decide if there is a solution or not.
In the next section we present the manipulation of a generalized equation by an example.

2.2 Transformation of a generalized equation

It is possible to transform generalized equations in such a way that we come up with an generalized equation which trivially can be decided. As an example we give some steps of the transformation of the generalized equation from the previous section.
At first we complete the partial order of the generalized equation obtaining a total order. Clearly there are finitely many ways to do this. The original generalized equation has a solution if one of the totally ordered generalized equations has a solution. For this example we choose the ordering

$$b_0 < b_5 < b_1 < b_2 < b_6 < b_3 < b_4 < b_7 < b_8.$$

The aim of the transformation is to reduce the length of the solution without increasing the complexity of the equation.
For the generalized equation described above $\sigma(x_1)$ must be of the form

$$\sigma(x_1) = as, \quad s \in S.$$

We want to reduce the length of $\sigma(x_1)$ by skipping this a which means that we solve a modified equation:

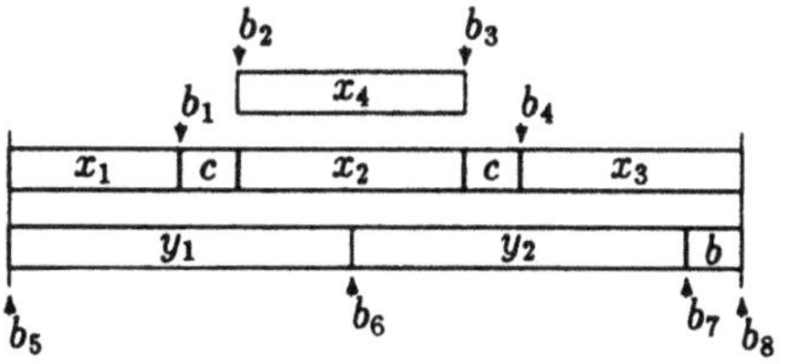

To obtain such a generalized equation we at first move a from x_1 to x_2. This does not change the solutions of the generalized equation since $\sigma(x_1) = \sigma(x_2)$. We generate a new boundary b_9, $b_2 < b_9 < b_3$, and get the following picture:

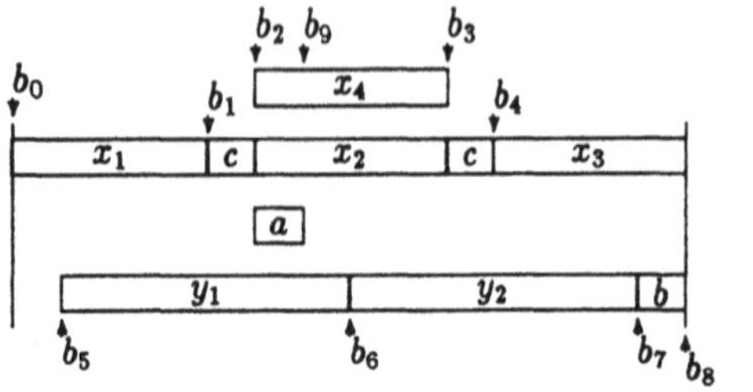

From the possible total orderings we have chosen $b_2 < b_9 < b_6$.
After moving a we do not know that the piece of x_1 between b_0 and b_5 is as long as the piece of x_2 between b_2 and b_9. To keep this information we add the boundary connection $(b_5/x_1/b_9)$ which means that b_5 divides x_1 at the same place where b_9 divides $\Delta x_1 = x_2$:

A solution σ must satisfy $\sigma(b_0)s = \sigma(b_5)$ and $\sigma(b_2)s = \sigma(b_9)$ for some $s \in S$.

Analogously to a we have to move the restriction $(b_0, \neg c, b_1)$ to keep the information that $c \not\ll \sigma(x_1)$. Thus we need the additional restriction $(b_2, \neg c, b_9)$.
Now we can skip the first part of x_1. Since $\sigma(x_1) = \sigma(x_2)$ we have also to skip the first part of x_2 and get the equation

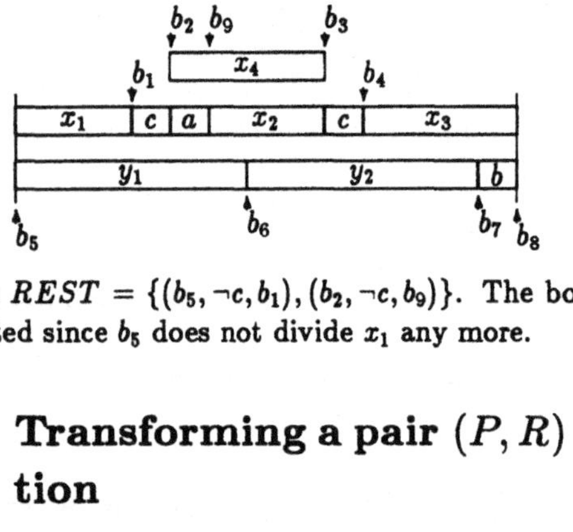

with $REST = \{(b_5, \neg c, b_1), (b_2, \neg c, b_9)\}$. The boundary connection (b_5, x_1, b_9) must be deleted since b_5 does not divide x_1 any more.

3 Transforming a pair (P, R) into a generalized equation

We want to find an algorithm with the following property.

Algorithm GENERATE_EQUATION$(P, R) = G$

Precondition

(a) P is a finite set of string equations.

(b) R is a corresponding set of constant restrictions.

1. Calculate G.

Postcondition

(a) G is a generalized equation.

(b) (P, R) has a solution $\Longrightarrow$ G has a short solution.

(c) (P, R) has no solution $\Longrightarrow G$ has no solution.

Clearly (P, R) has a solution if there is a substitution solving all equations in P and satisfying the constant restrictions of R.

3.1 Bounding the number of repetitions

At first we want to calculate the number REP. This is done by a theorem similar to a lemma of Makanin.

Theorem 2 (**[5, Lemma 1.3]**) *If a string equation $s \asymp t$ has a solution then it has a solution σ such that* $\mathrm{R}\#(\sigma, \mathcal{V}, \eta(|st|))$ *holds where*

$$\eta(m) = (6m)^{2^{2m^4}} + 2.$$

The proof of the theorem relies on the following lemma.

Lemma 1 *Let σ be a solution of the string equation $s \asymp t$ with*

$$\sigma(x) = v^{k_{x,0}} \prod_{i=1}^{n_x} w_{x,i} v^{k_{x,i}}$$

where $x \in \mathcal{V}$, $k_{x,i} \geq 0$, $v, w_{x,i} \in \mathcal{S}$. Then there exists a solution $\hat{\sigma}$ such that

$$\hat{\sigma}(x) = v^{\hat{k}_{x,0}} \prod_{i=1}^{n_x} w_{x,i} v^{\hat{k}_{x,i}}$$

with $\hat{k}_{x,i} \leq min\{k_{x,i}, \eta(|st|)\}$ for all $\hat{k}_{x,i}$.

The proof of the lemma is contained in the proof Makanin's lemma. For a worked out version see [2].
From Lemma 1 we easily get

Lemma 2 *If a string equation $s \asymp t$ has a solution which satisfies the restrictions in R then it has a solution σ such that the restrictions in R are satisfied and* $\mathrm{R}\#(\sigma, \mathcal{V}, \eta(|st|))$ *holds.*

Proof. Assume there is a solution σ which satisfies R with $v^k \ll \sigma(x)$ for some $v \in \mathcal{S}$, $x \in \mathcal{V}$, $k > \eta(|st|)$. Then by Lemma 1 there is a solution $\hat{\sigma}$ such that $v^k \not\ll \hat{\sigma}(x)$, $|\hat{\sigma}(y)| \leq |\sigma(y)|\ \forall y \in \mathcal{V}$, $|\hat{\sigma}(x)| < |\sigma(x)|$, and

$$a \not\ll \sigma(y) \Longrightarrow a \not\ll \hat{\sigma}(y) \quad \forall a \in \mathcal{C},\ \forall y \in \mathcal{V}.$$

Hence $\hat{\sigma}$ satisfies R, too.
Repeatedly applying Lemma 1 gives a solution of $s \asymp t$ which satisfies R and $\mathrm{R}\#(\sigma, \mathcal{V}, \eta(|st|))$.

□

Theorem 3 *Assume a pair (P, R) with $P = \{s_i \asymp t_i : 1 \leq i \leq n\}$. Define $m = \sum_{i=1}^{n} |s_i t_i|$ and $REP = \eta(m + 2n - 2)$. Then (P, R) has a solution if and only if it has a solution σ such that* $\mathrm{R}\#(\sigma, \mathcal{V}, REP)$ *holds.*

Proof. Let $c \in \mathcal{C}$ be a *new* constant with $c \notin \text{const}(P)$. Then σ solves (P, R) if and only if it solves $(\{s \asymp t\}, R')$ where

$$\begin{aligned} s &= s_1 c s_2 c \cdots c s_n, \\ t &= t_1 c t_2 c \cdots c t_n, \\ R' &= R \cup \{(c \not\ll x) : x \in \text{var}(P)\}. \end{aligned}$$

This is easily seen: Let σ be a solution of (P, R). Since $c \notin \text{const}(P)$ we have $c \not\ll \sigma(x)$ for all $x \in \text{var}(P)$. Hence σ satisfies R'. Furthermore

$$\sigma(s) = \sigma(s_1)c\sigma(s_2)c \cdots c\sigma(s_n) = \sigma(t_1)c\sigma(t_2)c \cdots c\sigma(t_n) = \sigma(t)$$

and σ solves $(\{s \asymp t\}, R')$, too.
If on the other hand σ solves $(\{s \asymp t\}, R')$ then σ satisfies R. Since $c \notin \text{const}(P)$ and $c \not\ll \sigma(x)$ for $x \in \text{var}(P)$ we have that $c \not\ll s_i$, $c \not\ll t_i$ for $i = 1, \ldots, n$. Hence $\sigma(s) = \sigma(t)$ implies $\sigma(s_i) = \sigma(t_i)$ for $i = 1, \ldots, n$ and σ solves (P, R).
Clearly $|st| = m + 2n - 2$ and therefore Lemma 2 proves the theorem.

□

Remark 4 *Theorem 3 gives the needed number REP.*

3.2 Generating the boundaries

Assume a pair (P, R) with

$$P = \{u_{i,1}u_{i,2} \cdots u_{i,m_i} \asymp v_{i,1}v_{i,2} \cdots v_{i,n_i} : 1 \le i \le k\}$$

where $u_{i,j}, v_{i,j} \in \mathcal{C} \cup \mathcal{V}$, $k \ge 0$ and $m_i, n_i \ge 1$. We construct a pair (P^B, R) with

$$P^B = \bigcup_{1 \le i \le k} \left(\{b_{i,j-1}u_{i,j} \asymp b_{i,j} : 1 \le j \le m_i\} \cup \{b'_{i,j-1}v_{i,j} \asymp b'_{i,j} : 1 \le j \le n_i\}\right)$$

where the $b_{i,j}, b'_{i,j}$ are new variables, i.e. they do not appear in P or R. Furthermore all variables $b_{i,j}, b'_{i,j}$ are distinct, except that

$$\begin{aligned} b_0 &= b_{1,0} = b'_{1,0}, \\ b_{i,0} &= b'_{i,0} = b_{i-1,m_i} = b'_{i-1,n_i} \quad \forall i \in \{2, \ldots, n\}. \end{aligned} \tag{1}$$

Then *BOUND* is given as the set

$$B = \{b_0\} \cup \bigcup_{1 \le i \le k} \left(\{b_{i,j} : 1 \le j \le m_i\} \cup \{b'_{i,j} : 1 \le j \le n_i - 1\}\right)$$

with the relations

$$b_{i,j-1} < b_{i,j} \quad \forall 1 \le j \le m_i \; \forall 1 \le i \le k$$

and

$$b'_{i,j-1} < b'_{i,j} \quad \forall 1 \le j \le n_i \; \forall 1 \le i \le k.$$

Remark 5 *Observe that all boundaries except b_0 are right boundaries.*

Example 2. Assume that $P = \{xx \asymp cyd, yy \asymp dxc\}$ and $R = \{(c \not\leq y)\}$ where $c, d \in \mathcal{C}$ and $x, y \in \mathcal{V}$. Then we construct P^B as

$$P^B = \{ \quad b_0x \asymp b_{1,1}, b_{1,1}x \asymp b_{1,2}, b_0c \asymp b'_{1,1}, b'_{1,1}y \asymp b'_{1,2}, b'_{1,2}d \asymp b_{1,2}, \\ b_{1,2}y \asymp b_{2,1}, b_{2,1}y \asymp b_{2,2}, b_{1,2}d \asymp b'_{2,1}, b'_{2,1}x \asymp b'_{2,2}, b'_{2,2}c \asymp b_{2,2}\}$$

or after renaming

$$P^B = \{ \quad b_0x \asymp b_1, b_1x \asymp b_2, b_0c \asymp b_3, b_3y \asymp b_4, b_4d \asymp b_2, \\ b_2y \asymp b_5, b_5y \asymp b_6, b_2d \asymp b_7, b_7x \asymp b_8, b_8c \asymp b_6\}.$$

□

Lemma 3 *(P, R) has a solution σ_1 if and only if (P^B, R) has a solution σ_2 which satisfies C_{BOUND}. Futhermore $\mathrm{R\#}(\sigma_1, \mathrm{var}(P), r)$ holds for some solution σ_1 of (P, R) if and only if $\mathrm{R\#}(\sigma_2, \mathrm{var}(P), r)$ holds for some solution σ_2 of (P^B, R).*

Proof. Assume that (P, R) has a solution σ_1. We define the substitution σ_2 as

$$\sigma_2(x) = \sigma_1(x) \text{ for all } x \in \mathcal{V} - B,$$

$$\sigma_2(b_0) = b_0,$$

$$\sigma_2(b_{i,j}) = \sigma_2(b_{i,j-1})\sigma_2(u_{i,j}) \text{ for } 1 \leq j \leq m_i,$$

$$\sigma_2(b'_{i,j}) = \sigma_2(b'_{i,j-1})\sigma_2(v_{i,j}) \text{ for } 1 \leq j \leq n_i - 1.$$

Clearly σ_2 satisfies the restrictions in R, $\mathrm{R\#}(\sigma_1, \mathrm{var}(P), r)$ implies $\mathrm{R\#}(\sigma_2, \mathrm{var}(P), r)$, and by the definition of σ_2 C_{BOUND} holds. We only have to prove that σ_2 solves P^B. By definition of σ_2 we have

$$\sigma_2(b_{i,j-1}u_{i,j}) = \sigma_2(b_{i,j-1})\sigma_2(u_{i,j}) = \sigma_2(b_{i,j}) \text{ for } 1 \leq j \leq m_i,$$

$$\sigma_2(b'_{i,j-1}v_{i,j}) = \sigma_2(b'_{i,j}) \text{ for } 1 \leq j \leq n_i - 1.$$

At last we must show that $\sigma_2(b'_{i,n_i-1}v_{i,n_i}) = \sigma_2(b'_{i,n_i})$. It can easily be seen by induction that

$$\sigma_2(b_{i,m_i}) = \sigma_2(b_{i,0})\sigma_1(u_{i,1} \cdots u_{i,m_i}), \\ \sigma_2(b'_{i,n_i-1}) = \sigma_2(b_{i,0})\sigma_1(v_{i,1} \cdots v_{i,n_i-1}).$$

Since $\sigma_1(u_{i,1} \cdots u_{i,m_i}) = \sigma_1(v_{i,1} \cdots v_{i,n_i})$ we have

$$\sigma_2(b'_{i,n_i-1}v_{i,n_i}) = \sigma_2(b_{i,0})\sigma_1(v_{i,1} \cdots v_{i,n_i}) = \sigma_2(b_{i,0})\sigma_1(u_{i,1} \cdots u_{i,m_i}) = \sigma_2(b'_{i,n_i})$$

because $b_{i,m_i} = b'_{i,n_i}$.
Now assume that σ_2 is a solution of (P^B, R). By induction we easily get

$$\sigma_2(b_{i,m_i}) = \sigma_2(b_{i,0})\sigma_2(u_{i,1} \cdots u_{i,m_i}) \quad \text{and} \quad \sigma_2(b'_{i,n_i}) = \sigma_2(b'_{i,0})\sigma_2(v_{i,1} \cdots v_{i,n_i}).$$

Since $b_{i,m_i} = b'_{i,n_i}$ and $b_{i,0} = b'_{i,0}$ this implies that σ_2 is a solution of (P, R), too.

□

3.3 Generating the base equations

Since in the set of base equations each variable $x \in V$ must appear exactly once, we have to replace multiple occurences of x by distinct variables $x_1, x_2, \ldots$
The set P^B from the previous section can be written as the union of disjoint sets

$$P^B = P_0 \cup \bigcup_{x \in \mathcal{V}-B} P_x$$

where

$$P_0 = \{ba \asymp b' \in P^B : a \in \mathcal{C}\},$$
$$P_x = \{bx \asymp b' \in P^B\}$$

($P_x \neq \emptyset$ only for finitely many x). For all

$$P_x = \{b_i^x x \asymp b_i'^x : 1 \leq i \leq k_x\},\ k_x \geq 1,$$

we construct

$$P_x^1 = \{b_i^x x_i \asymp b_i'^x : 1 \leq i \leq k_x\} \cup \{b_1^x x_{i+k_x} \asymp b_1'^x : 1 \leq i \leq k_x\}$$

where $x_1 = x$ and $x_2, x_3, \ldots, x_{2k_x} \in \mathcal{V} - B$ are new distinct variables such that for all P_x^1, P_y^1 we have $x_i \neq y_j$ if $x \neq y$ or $i \neq j$.
The set of base equations $BASE$ is then given as

$$BASE = P_0 \cup \bigcup_{x \in \mathcal{V}-B, P_x \neq \emptyset} P_x^1$$

and VAR is given as

$$V = \bigcup_{x \in \mathcal{V}-B, P_x \neq \emptyset} \{x_i : 1 \leq i \leq 2k_x\}$$

with

$$\Delta x_i = \begin{cases} x_{i+k_x} & 1 \leq i \leq k_x \\ x_{i-k_x} & k_x + 1 \leq i \leq 2k_x. \end{cases}$$

Remark 6 *Observe that again all boundaries except b_0 are right boundaries since only variables have been replaced.*

Example 3. From the example of the previous section we get

$$P_0 = \{b_0 c \asymp b_3, b_4 d \asymp b_2, b_2 d \asymp b_7, b_8 c \asymp b_6\},$$
$$P_x = \{b_0 x \asymp b_1, b_1 x \asymp b_2, b_7 x \asymp b_8\},$$
$$P_y = \{b_3 y \asymp b_4, b_2 y \asymp b_5, b_5 y \asymp b_6\},$$

$$P_x^1 = \{b_0 x \asymp b_1, b_1 x_2 \asymp b_2, b_7 x_3 \asymp b_8, b_0 x_4 \asymp b_1, b_0 x_5 \asymp b_1, b_0 x_6 \asymp b_1\},$$
$$P_y^1 = \{b_3 y \asymp b_4, b_2 y_2 \asymp b_5, b_5 y_3 \asymp b_6, b_3 y_4 \asymp b_4, b_3 y_5 \asymp b_4, b_3 y_6 \asymp b_4\}$$

with $\Delta x = x_4$, $\Delta x_2 = x_5$, $\Delta x_3 = x_6$, $\Delta y = y_4$, $\Delta y_2 = y_5$, $\Delta y_3 = y_6$.

□

Lemma 4 *(P^B, R) has a solution σ_1 which satisfies C_{BOUND} if and only if there is a substitution σ_2 which satisfies $C_{BOUND}, C_{VAR}, C_{BASE}$ and all restrictions in R. Futhermore $\mathrm{R\#}(\sigma_1, V, r)$ holds for some σ_1 if and only if $\mathrm{R\#}(\sigma_2, V, r)$ holds for some σ_2.*

Proof. Let σ_1 be a solution of (P^B, R) which satisfies C_{BOUND}. We define a substitution σ_2 by

$$\sigma_2(y) = \sigma_1(y) \quad \forall y \in \mathcal{V} - V$$
$$\sigma_2(x_i) = \sigma_1(x) \quad \forall x_i \in V.$$

Clearly $\mathrm{R\#}(\sigma_1, \mathcal{V}-B, r)$ implies $\mathrm{R\#}(\sigma_2, \mathcal{V}-B, r)$ and C_{BOUND} holds for σ_2. Furthermore

$$\sigma_2(b_i^x x_i) = \sigma_2(b_i^x)\sigma_2(x_i) = \sigma_1(b_i^x)\sigma_1(x) = \sigma_1(b_i'^x) = \sigma_2(b_i'^x),$$
$$\sigma_2(b_1^x x_i) = \sigma_1(b_1^x x) = \sigma_1(b_1'^x) = \sigma_2(b_1'^x)$$

and σ_2 satisfies C_{BASE}. Since $\sigma_2(x_i) = \sigma_1(x) = \sigma_2(x_{i \pm k_x})$ σ_2 satisfies C_{VAR} and all restrictions in R.
Let σ_2 satisfy $C_{BOUND}, C_{VAR}, C_{BASE}$ and all restrictions in R. We show that $\sigma_2(x_i) = \sigma_2(x)$ for all $x_i \in V$.
For all x_i, $i > k_x$, we have $\sigma_2(b_1^x x_i) = \sigma_2(b_1'^x)$. Since $\sigma_2(b_1^x x_1) = \sigma_2(b_1'^x)$ this implies $\sigma_2(x_i) = \sigma_2(x_1) = \sigma_2(x)$ $(x = x_1)$.
For $i \leq k_x$ it follows that $\sigma_2(b_1^x x_i) = \sigma_2(b_1^x \Delta x_i) = \sigma_2(b_1'^x) = \sigma_2(b_1^x x_1)$ and hence again $\sigma_2(x_i) = \sigma_2(x)$.
Therefore $\sigma_2(b_i^x x) = \sigma_2(b_i^x x_i) = \sigma_2(b_i'^x)$ and σ_2 solves P^B, too.

□

3.4 Generating the modified restrictions

Lemma 5 *Let BOUND, VAR, BASE, and a set of restrictions R be given. Define*

$$REST = \{(\beta_l(x), \neg a, \beta_r(x)) : a \in R(x)\}.$$

Then σ is a solution of the generalized equation $(BOUND, VAR, BASE, \emptyset, REST, r)$ if and only if σ satisfies $C_{BOUND}, C_{VAR}, C_{BASE}$ and all restrictions in R. Furthermore σ is a short solution if and only if it also satisfies $\mathrm{R\#}(\sigma, V, r)$.

Proof. Since σ solves $(BOUND, VAR, BASE, \emptyset, REST, r)$ if and only if σ satisfies $C_{BOUND}, C_{VAR}, C_{BASE}$, and C_{REST}, we must show that σ satisfies $C_{REST} \Longleftrightarrow \sigma$ satisfies all restrictions in R.
Let $a \in R(x)$ and let $(\beta_l(x), \neg a, \beta_r(x))$ be the corresponding restriction in $REST$. Since $\beta_l(x)x \asymp \beta_r(x) \in BASE$ we have by C_{BASE} that $\sigma(\beta_l(x))\sigma(x) = \sigma(\beta_r(x))$. Hence $a \not\preccurlyeq \sigma(x)$ if and only if σ satisfies $(\beta_l(x), \neg a, \beta_r(x))$.

□

Example 4. If we consider the example from the previous section with $R(y) = \{c\}$ we get $REST = \{(b_3, \neg c, b_4)\}$.

□

3.5 Summing up

Clearly the steps of sections 3.1–3.4 can be formulated as algorithm.

Algorithm GENERATE_EQUATION$(P, R) = G$

Precondition

(a) P is a finite set of string equations.

(b) R is a set of corresponding restrictions.

1. Perform the steps of sections 3.1–3.4.

Postcondition

(a) G is a generalized equation.

(b) (P, R) has a solution $\Longrightarrow$ G has a short solution.

(c) (P, R) has no solution $\Longrightarrow$ G has no solution.

(d) All boundaries except the minimal boundary b_0 are right boundaries.

(e) The set of boundary connections is empty.

Proof. Theorem 3, Lemma 3, Lemma 4, and Lemma 5 prove the postcondition.

□

Remark 7 *Joining Theorem 3, Lemma 3, Lemma 4 we have to notice that $\mathcal{V}$ can be replaced by* var(P) *and* var(P) *can be replaced by V.*

4 The structure of the decision algorithm

In this section we present the structure of the algorithm which will decide whether a generalized equation has a (short) solution or not. We present the algorithm in a quite abstract way; the following sections will show how this abstract algorithm applies to our problem.

Let Ω be some set with a computable function $f : \Omega \to \mathbf{N} = \{0, 1, 2, \ldots\}$ such that $\{\omega : f(\omega) \leq n\}$ is finite for all $n \in \mathbf{N}$. Furthermore let $\Lambda_0 \subseteq \Lambda \subseteq \Omega$. We construct an algorithm DECIDE_$\Omega : \Omega \to \{\text{TRUE}, \text{FALSE}\}$ such that

$$\begin{aligned} \text{DECIDE\_}\Omega(\omega) = \text{TRUE} \quad &\Longrightarrow \quad \omega \in \Lambda \\ \text{DECIDE\_}\Omega(\omega) = \text{FALSE} \quad &\Longrightarrow \quad \omega \notin \Lambda_0. \end{aligned}$$

Remark 8 *We intend Ω to be the set of generalized equations, Λ the set of solvable equations, Λ_0 the set of equations with short solutions.*

We use a transformation algorithm $T : \Omega \to \mathcal{P}(\Omega)$ and a function $g : \Lambda \to \mathbf{N}$ with the following properties:

P1: $T(\omega)$ is finite for all $\omega \in \Omega$.

P2: $\omega_1 \in T(\omega) \Longrightarrow f(\omega_1) \leq f(\omega)$.

P3: $T(\omega) \cap \Lambda \neq \emptyset \Longrightarrow \omega \in \Lambda$.

P4: $\omega \in \Lambda_0, f(\omega) \neq 0 \Longrightarrow \exists \omega_1 \in T(\omega) \cap \Lambda_0 : g(\omega_1) < g(\omega)$.

P5: $f(\omega) = 0 \Longrightarrow \omega \in \Lambda$.

Then we have

Algorithm DECIDE_$\Omega(\omega) \in \{\text{TRUE}, \text{FALSE}\}$

Precondition

(a) $\omega \in \Omega$.

1. $P{:=}\emptyset$; $O{:=}\{\omega\}$.

2. LOOP
   IF $\exists o \in O : f(o) = 0$ THEN RETURN (TRUE)
   ELSEIF $O = \emptyset$ THEN RETURN (FALSE)
   ELSE $P{:=}P \cup O$; $O{:=}\bigcup_{o \in O} T(o) - P$
   ENDLOOP .

Postcondition

(a) DECIDE_$\Omega(\omega) = \text{TRUE} \Longrightarrow \omega \in \Lambda$.
(b) DECIDE_$\Omega(\omega) = \text{FALSE} \Longrightarrow \omega \notin \Lambda_0$.

4..1 Correctness proof for algorithm DECIDE_Ω

To prove the correctness of the algorithm we use the invariant $I = I_1 \wedge I_2 \wedge I_3 \wedge I_4 \wedge I_5 \wedge I_6$

I_1 : $P \cup O$ is finite.

$I_2 : P \cap O = \emptyset$.

$I_3 : o \in P \cup O \Longrightarrow f(o) \leq f(\omega)$.

$I_4 : O \cap \Lambda \neq \emptyset \Longrightarrow \omega \in \Lambda$.

$I_5 : \omega \in \Lambda_0 \Longrightarrow O \cap \Lambda_0 \neq \emptyset$.

$I_6 : \min\{g(o) : o \in O \cap \Lambda_0\} < \min\{g(p) : p \in P \cap \Lambda_0\}$.

In I_6 we set $\min N < \min \emptyset$ for all $N \subseteq \mathbf{N}$.
Clearly I holds after step 1 of the algorithm. If $o \in O$ with $f(o) = 0$ then by P5 $o \in O \cap \Lambda$ and I_4 implies $\omega \in \Lambda$. If $O = \emptyset$ then $O \cap \Lambda_0 = \emptyset$ and I_5 implies $\omega \notin \Lambda_0$.
Now we assume $f(o) \neq 0$ for all $o \in O$. Let $P_1 = P \cup O$ and $O_1 = T(O) - P - O$ (for shortness we write $T(O)$ instead of $\bigcup_{o \in O} T(o)$). We have to prove that I holds for P_1, O_1 if it holds for P, O.

1. Obviously $P_1 \cup O_1$ is finite by P1.

2. $O_1 = T(O) - P_1$ implies $P_1 \cap O_1 = \emptyset$.

3. Since $o_1 \in P_1 \cup O_1$ implies $o_1 \in P \cup O$ or $o_1 \in T(O)$, by P2 $f(o_1) \leq f(\omega)$.

4. By P3 we have $O_1 \cap \Lambda \neq \emptyset \Rightarrow T(O) \cap \Lambda \neq \emptyset \Rightarrow O \cap \Lambda \neq \emptyset \Rightarrow \omega \in \Lambda$.

5. Let $m \in O \cap \Lambda_0$ with $g(m) = \min\{g(o) : o \in O \cap \Lambda_0\}$, $f(m) \neq 0$. Then by P4 there is an $m_1 \in T(O) \cap \Lambda_0$ with $g(m_1) < g(m)$. Clearly $m_1 \notin P \cup O$ (since I_6) and hence $m_1 \in O_1$.

6. By the proof of I_5 we have $m_1 \in O_1 \cap \Lambda_0$, $g(m_1) < \min\{g(p) : p \in P_1 \cap \Lambda_0\}$ or $O \cap \Lambda_0 = \emptyset$ which implies $P_1 \cap \Lambda_0 = \emptyset$.

Since $P \subset P \cup O = P_1$ if $O \neq \emptyset$ (remember $P \cap O = \emptyset$) and $P_{max} = \{o \in \Omega : f(o) \leq f(\omega)\}$ is finite we have $|P_{max} - P_1| < |P_{max} - P|$. Hence the algorithm terminates.

5 How to get Ω, Λ, Λ_0, f

To apply the decision algorithm of the previous section we have to define the sets Ω, Λ, Λ_0 and the computable function $f : \Omega \to \mathbf{N}$ and the transformation algorithm T. Clearly we want Ω to be the set of all generalized equations. But unfortunately this set is too big to be dealt with reasonably. Hence we restrict Ω to the subset of *totally proper* generalized equations.

Definition 7 (Proper generalized equation) *A generalized equation is called* proper *if it satisfies the following conditions:*

PROP1: *All boundaries except the minimal boundary b_0 are right boundaries or appear in a boundary connection.*

PROP2: *The boundary connections are related by the binary relation $\trianglelefteq$ which is given below, such that for all $con \in CONN$ there exists $con_0, con_r = (b, x, b_r) \in CONN$ with*

$$con_0 \trianglelefteq con, \; con_0 \trianglelefteq con_r$$

and

$$b_r = \beta_r(y), \beta_l(x) \leq \beta_l(y) \;\; or \;\; b_r\epsilon \asymp b_r \in BASE \;\; or \;\; con_0 = con_r.$$

Definition 8 *The relation $\preceq = \preceq_G$ on boundary connections is defined as follows*

$$(b_1/x/b_2) \preceq (b_2/y/b_3) \Leftrightarrow \beta_l(\Delta x) \leq \beta_l(y).$$

Then $\trianglelefteq = \trianglelefteq_G$ is defined by

$$con_a \trianglelefteq con_b \Leftrightarrow \exists con_1, \ldots, con_k : con_a \preceq con_1 \preceq \cdots \preceq con_k \preceq con_b.$$

Definition 9 (Totally proper generalized equation) *A proper generalized equation is called* totally proper *if it satisfies the following conditions:*

TP1: *The set of boundaries is totally ordered.*

TP2: *The names of variables and constants are fixed such that*

$$B = \{b_0, b_1, \ldots, b_M\}$$
$$V = \{x_1, x_2, \ldots, x_N\}$$
$$\text{const}(BASE) = \{c_1, c_2, \ldots, c_K\}$$

for some numbers M, N, K.

TP3: *The number of boundary connections is bounded by a computable function $\eta \geq 0$ which will be given later:*

$$|CONN| \leq \eta(|BASE|, REP)$$

Now Ω is the set of all totally proper generalized equations and $\Lambda \subseteq \Omega$ is the set of solvable equations and $\Lambda_0 \subseteq \Lambda$ is the set of equations with short solutions.
The function $f : \Omega \to \mathbf{N}$ is defined by

$$f(BOUND, VAR, BASE, CONN, REST, REP) = |BASE| * REP.$$

Since $|BASE| = 0$ implies $|VAR| = 0, |CONN| = 0, |REST| = 0$ and hence $B = \{b_0\}$ (by PROP1), all substitutions are solutions of a proper generalized equation G with $f(G) = 0$.
What remains to prove is that $\{G \in \Omega : f(G) \leq n\}$ is finite for all $n \in \mathbf{N}$.

Theorem 4 *There are only finitely many totally proper generalized equations*

$$(BOUND, VAR, BASE, CONN, REST, REP)$$

with

$$|BASE| * REP \leq n.$$

Proof. Clearly

$$M = |B| - 1 \leq |BASE| + |CONN| \leq |BASE| + \eta(|BASE|, REP)$$
$$N \leq |BASE|$$
$$K \leq |BASE|.$$

If $|BASE| = 0$ then $|M| = 1$. Hence we are finished since there are only finitely many ways to build $(BOUND, VAR, BASE, CONN, REST, REP)$.

□

6 How to obtain totally proper equations

To apply DECIDE_Ω we have to transform the (proper) generalized equation of section 3 into a totally proper generalized equation:

Algorithm TOTAL$(G) = \mathcal{G}$

Precondition

(a) $G = (BOUND, VAR, BASE, CONN, REST, REP)$ is a proper generalized equation.

1. Complete the partial order of $BOUND$ in all possible ways obtaining equations $G_i = (BOUND_i, VAR_i, BASE_i, CONN_i, REST_i, REP)$, $i = 1, \ldots, n$.
2. Rename the variables and constants of equations $G_1, \ldots, G_n$ accordingly to theorem 4.
3. <u>RETURN</u> $(\{G_i : |CONN_i| \leq \eta(|BASE_i|, REP)\})$.

Postcondition

(a) $\mathcal{G}$ is a finite set of totally proper generalized equations.

(b) G has a short solution $\Longrightarrow$ some $G' \in \mathcal{G}$ has a short solution.

(c) Some $G' \in \mathcal{G}$ has a solution $\Longrightarrow$ G has a solution.

(d) $G' \in \mathcal{G} \Rightarrow f(G') \leq f(G)$.

Remark 9 *Boundaries might be unified by step 1 of the algorithm.*

Proof.

1. Clearly there are only finitely many ways to complete the partial order of $BOUND$ and this step is quite trivial. Since a (short) solution σ of G yields a total ordering of the boundaries σ must be (short) solution of some G_i, too. And trivially any solution of some G_i is solution of G, too. Furthermore $f(G_i) \leq f(G)$ for REP is not changed and no base equations are added.
2. The renaming does not make any significant changes.
3. The function η can be chosen such that proper generalized equations with $|CONN| > \eta(|BASE|, REP)$ have no short solutions. See the next section.

□

Remark 10 *Finally, instead of one generalized equation we get a set of totally proper generalized equations such that the original equation has a solution if an equation in the set has a solution. This question can be answered by algorithm* DECIDE_Ω *applied to all equations in the set.*

6.1 The function $\eta(|BASE|, REP)$

We have to calculate η such that a proper generalized equation has no solution if $|CONN| > \eta(|BASE|, REP)$. This can be done be the Main Lemma of [4, p.75]. For a worked out proof see [2].

7 The transformation

At last we have to find an algorithm $T : \Omega \to \mathcal{P}(\Omega)$ and a function $g : \Lambda \to \mathbf{N}$ with properties P1–P5. Because of algorithm TOTAL we only need an algorithm $T_0 : \Omega \to \Omega^+ \cup \{\text{FALSE}\}$ where Ω^+ is the set of all proper generalized equations and $T_0(G) =$ FALSE implies that G has no solution.
As indicated in section 2.2 there are two main cases for the transformation of a generalized equation:

- Moving a variable or constant to another position.
- Skipping the first part of an equation.

This description is very informal and we will give a formal one later. Anyhow we can see that the transformation either moves some part of the equation from left to right or skips some part of the equation. Since the transformation must decrease function g we have to choose a function which is decreased by skipping parts or moving parts from left to right:
Let σ be a solution of the generalized equation

$$G = (BOUND, VAR, BASE, CONN, REST, REP)$$

. We define

$$g_0(G, \sigma) = \sum_{b_l x \asymp b_r \in BASE} 2^{|\sigma(b_M)| - |\sigma(b_l)|}$$

where $|\sigma(b_M)| = \max\{|\sigma(b)| : b \in B\}$. Then we have

$$g(G) = \min\{g_0(G, \sigma) : \sigma \text{ solves } G\}.$$

It is not very hard to see that g is decreased by the transformations mentioned above since moving from left to right increases the length of $\sigma(b)$. Precise argumentation will be given while dealing with the transformation in detail.

Lemma 6 *Let G be a proper generalized equation with a short solution σ. Then some equation $G' \in$ TOTAL(G) has a short solution $\hat{\sigma}$ with $g_0(G', \hat{\sigma}) \leq g_0(G, \sigma)$.*

Proof. By the proof for algorithm TOTAL and the definition of g_0.

□

Hence we need an

Algorithm $T_0(G) \in \Omega^+ \cup \{\text{FALSE}\}$

Precondition

(a) G is a totally proper equation with $f(G) > 0$.

1. Calculate $G_T \in \Omega^+$ or return FALSE.

Postcondition

(a) $T_0(G) = \text{FALSE} \Longrightarrow G$ has no solution.

(b) $f(G_T) \leq f(G)$.

(c) G has a short solution $\sigma \Longrightarrow G_T$ has a short solution $\hat{\sigma}$ with $g_0(G_T, \hat{\sigma}) < g_0(G, \sigma)$.

(d) G_T has a solution $\Longrightarrow G$ has a solution.

7.1 Case analysis

As mentioned earlier the transformation has to deal with some cases which are defined in the following:

C1. There exists $x_0 \in V$ with $\beta_l(x_0) = b_0, \beta_l(\Delta x_0) \neq b_0$ and there exists $b_0 x \asymp b_r \in BASE$ with $x \neq x_0$, $b_0 < b_r \leq \beta_r(x_0)$.

C2. There exists $x_0 \in V$ with $\beta_l(x_0) = b_0, \beta_l(\Delta x_0) \neq b_0$ and there exists no $b_0 x \asymp b_r \in BASE$ with $x \neq x_0$.

C3. For all $x_0 \in V$ with $\beta_l(x_0) = b_0, \beta_l(\Delta x_0) \neq b_0$ there exists $b_0 x \asymp b_r \in BASE$ with $x \neq x_0, b_r > \beta_r(x_0)$ or $x = \epsilon$.

Remark 11 *In case C1 one part of the equation has to be moved. In case C2 the first part of the equation has to be skipped. Case C3 is the complement of C1,C2 and contains some special cases which must be dealt with seperately.*

7.2 Notation of the transformation

For the transformations given in the following we assume that a totally proper generalized equation $G = (BOUND, VAR, BASE, CONN, REST, REP)$ with $f(G) \neq 0$ is given. For notational simplicity let the boundary connections be numbered, $CONN = \{con_1, con_2, \ldots, con_{|CONN|}\}$.
Then the transformation produces an equation

$$G_T = (BOUND_T, VAR_T, BASE_T, CONN_T, REST_T, REP) \in \Omega^+$$

(REP is never changed) with the properties given above. If G trivially has no solution then the transformation returns FALSE.
During the transformation new boundaries might by introduced which we will denote by b^*. *New* means that $b^* \notin B$.

8 Transformation of case C1

Let $x_0 \in V$ with $\beta_l(x_0) = b_0, \beta_l(\Delta x_0) \neq b_0$ and $b_0 x \asymp b_r \in BASE$ with $x \neq x_0$ $b_0 < b_r \leq \beta_r(x_0)$. We distinguish two subcases $b_0 < b_r < \beta_r(x_0)$ and $b_r = \beta_r(x_0)$ which are quite similar.

8.1 Subcase 1: $b_0 < b_r < \beta_r(x_0)$

The transformation consists of the following:
Let b_r^* be a new boundary. Then

$$
\begin{aligned}
BASE_T &:= BASE - \{b_0 x \asymp b_r\} + \{\beta_l(\Delta x_0)x \asymp b_r^*\} \\
CONN_T &:= CONN + \{(b_r/x_0/b_r^*)\} \\
&\quad -\{con_i : con_i = (b_{i1}/x/b_{i2})\} \\
&\quad +\{(b_{i1}/x_0/b_i^*), (b_i^*/x/b_{i2}) : con_i = (b_{i1}/x/b_{i2})\} \\
BOUND_T &: B_T := B + \{b_r^*\} + \{b_i^* : con_i = (b_{i1}/x/b_{i2})\} \\
&\quad \beta_l(\Delta x_0) < b_r^* < \beta_r(\Delta x_0), \beta_l(\Delta x_0) < b_i^* < b_r^*. \\
VAR_T &:= VAR \\
REST_T &:= REST
\end{aligned}
$$

8.2 Proof for subcase 1

Scanning the definitions it is easily seen that G_T is a generalized equation which satisfies PROP1 and that $f(G_T) \leq f(G)$.

8.2.1 G_T satisfies PROP2

If $(b_1/y/b_r) \in CONN$ and $\beta_l(\Delta y) \leq \beta_l(x)$ then $(b_1/y/b_r) \trianglelefteq (b_r/x_0/b_r^*)$.
We define a function $\pi : CONN \to CONN_T \cup CONN_T \times CONN_T$ as

$$
\begin{aligned}
&\pi(con) = con \quad \forall con \in CONN \cap CONN_T \\
&\pi(b_{i1}/x/b_{i2}) = (b_{i1}/x_0/b_i^*), (b_i^*/x/b_{i2}).
\end{aligned}
$$

It is easy to see that $con_1 \preceq_G con_2$ implies $\pi(con_1) \trianglelefteq_{G_T} \pi(con_2)$ where $con_1 \trianglelefteq_{G_T} con_2, con_3$ and $con_1, con_2 \trianglelefteq_{G_T} con_3$ mean $con_1 \trianglelefteq_{G_T} con_2 \trianglelefteq_{G_T} con_3$. Hence PROP2 is satisfied.

8.2.2 Let σ be a short solution of G

We construct a short solution $\hat{\sigma}$ of G_T:

$$
\begin{aligned}
\hat{\sigma}(x) &= \sigma(x) \quad \forall x \in V \\
\hat{\sigma}(b) &= \sigma(b) \quad \forall b \in B \\
\hat{\sigma}(b_r^*) &= \sigma(\beta_l(\Delta x_0)x) \\
\hat{\sigma}(b_i^*) &= \sigma(\beta_l(\Delta x_0))s_i \quad \text{with} \quad \sigma(b_0)s_i = \sigma(b_{i1}).
\end{aligned}
$$

- By $b_0 < b_r < \beta_r(x_0)$, $(b_{i1}/x/b_{i2})$ we have $\sigma(x)t = \sigma(x_0)$, $\sigma(b_0)s_it_i = \sigma(b_{i1})t_i = \sigma(b_r)$ for some s_i, t, t_i and hence

$$\hat{\sigma}(\beta_l(\Delta x_0))\sigma(x)t = \hat{\sigma}(b_r^*)t = \hat{\sigma}(\beta_r(\Delta(x_0))$$
$$\hat{\sigma}(\beta_l(\Delta x_0))s_it_i = \hat{\sigma}(b_i^*)t_i = \hat{\sigma}(\beta_l(\Delta(x_0)x) = \hat{\sigma}(b_r^*).$$

- Clearly $\hat{\sigma}(y) = \hat{\sigma}(\Delta y)$ for all $y \in V$.
- Clearly $\hat{\sigma}$ solves all base equations since $\hat{\sigma}(b_r^*) = \hat{\sigma}(\beta_l(\Delta x_0)x)$.
- We have

$$\hat{\sigma}(b_0)\sigma(x) = \hat{\sigma}(b_r),\ \hat{\sigma}(\beta_l(\Delta x_0))\sigma(x) = \hat{\sigma}(b_r^*)$$
$$\hat{\sigma}(b_0)s_i = \hat{\sigma}(b_{i1}),\ \hat{\sigma}(\beta_l(\Delta x_0))s_i = \hat{\sigma}(b_i^*),\ \hat{\sigma}(\beta_l(\Delta x))s_i = \hat{\sigma}(b_{i2}).$$

- Clearly all restrictions in $REST_T$ are satisfied.
- Clearly $R\#(\hat{\sigma}, V, REP)$ holds.

Since $b_0 < \beta_l(\Delta x_0)$ clearly $g_0(G_T, \hat{\sigma}) < g_0(G, \sigma)$.

8.2.3 Let σ be solution of G_T

We prove that σ is a solution of G, too.

- Clearly the relations in $BOUND$ are satisfied.
- Clearly $\sigma(y) = \sigma(\Delta y)$ for all $y \in V$.
- Since $\sigma(\beta_l(\Delta x_0))\sigma(x) = \sigma(b_r^*)$ we have $\sigma(b_0)\sigma(x) = \sigma(b_r)$ via $(b_r/x_0/b_r^*)$.
- Since $\sigma(b_0)s_i = \sigma(b_{i1}), \sigma(\beta_l(\Delta x_0))s_i = \sigma(b_i^*), \sigma(\beta_l(\Delta x))s_i = \sigma(b_{i2})$ σ satisfies $(b_{i1}/x/b_{i2})$ in G.
- Clearly all restrictions in $REST$ are satisfied.

8.3 Subcase 2: $b_r = \beta_r(x_0)$

The transformation is very similar to subcase 1. The main difference is that we do not introduce a new variable b_r^* but use $\beta_r(\Delta x_0)$ instead:

$$\begin{aligned}
BASE_T &:= BASE - \{b_0x \asymp b_r\} + \{\beta_l(\Delta x_0)x \asymp \beta_r(\Delta x_0)\}\\
CONN_T &:= CONN\\
&\quad -\{con_i : con_i = (b_{i1}/x/b_{i2})\}\\
&\quad +\{(b_{i1}/x_0/b_i^*), (b_i^*/x/b_{i2}) : con_i = (b_{i1}/x/b_{i2})\}\\
BOUND_T &: B_T := B + \{b_i^* : con_i = (b_{i1}/x/b_{i2})\}\\
&\quad \beta_l(\Delta x_0) < b_i^* < \beta_r(\Delta x_0).\\
VAR_T &:= VAR\\
REST_T &:= REST
\end{aligned}$$

8.4 Proof for subcase 2

Scanning the definitions it is easily seen that G_T is a generalized equation which satisfies PROP1 and that $f(G_T) \leq f(G)$.

8.4.1 G_T satisfies PROP2

We define a function $\pi : CONN \to CONN_T \cup CONN_T \times CONN_T$ as

$$\pi(con) = con \quad \forall con \in CONN \cap CONN_T$$
$$\pi(b_{i1}/x/b_{i2}) = (b_{i1}/x_0/b_i^*), (b_i^*/x/b_{i2}).$$

It is easy to see that $con_1 \preceq_G con_2$ implies $\pi(con_1) \trianglelefteq_{G_T} \pi(con_2)$ where $con_1 \trianglelefteq_{G_T} con_2, con_3$ and $con_1, con_2 \trianglelefteq_{G_T} con_3$ mean $con_1 \trianglelefteq_{G_T} con_2 \trianglelefteq_{G_T} con_3$. Hence PROP2 is satisfied.

8.4.2 Let σ be a short solution of G

We construct a short solution $\hat{\sigma}$ of G_T:

$$\hat{\sigma}(x) = \sigma(x) \quad \forall x \in V$$
$$\hat{\sigma}(b) = \sigma(b) \quad \forall b \in B$$
$$\hat{\sigma}(b_i^*) = \sigma(\beta_l(\Delta x_0))s_i \quad \text{with} \quad \sigma(b_0)s_i = \sigma(b_{i1}).$$

- By $b_r = \beta_r(x_0)$, $(b_{i1}/x/b_{i2})$ we have $\sigma(x) = \sigma(x_0)$, $\sigma(b_0)s_it_i = \sigma(b_{i1})t_i = \sigma(b_r)$ for some s_i, t_i and hence

$$\hat{\sigma}(\beta_l(\Delta x_0))s_it_i = \hat{\sigma}(b_i^*)t_i = \hat{\sigma}(\beta_l(\Delta(x_0)x) = \hat{\sigma}(\beta_r(\Delta x_0))$$

- Clearly $\hat{\sigma}(y) = \hat{\sigma}(\Delta y)$ for all $y \in V$.
- Clearly $\hat{\sigma}$ solves all base equations since $\sigma(x) = \sigma(x_0)$.
- We have

$$\hat{\sigma}(b_0)s_i = \hat{\sigma}(b_{i1}),\ \hat{\sigma}(\beta_l(\Delta x_0))s_i = \hat{\sigma}(b_i^*),\ \hat{\sigma}(\beta_l(\Delta x))s_i = \hat{\sigma}(b_{i2}).$$

- Clearly all restrictions in $REST_T$ are satisfied.
- Clearly R#($\hat{\sigma}, V, REP$) holds.

Since $b_0 < \beta_l(\Delta x_0)$ clearly $g_0(G_T, \hat{\sigma}) < g_0(G, \sigma)$.

8.4.3 Let σ be solution of G_T

We prove that σ is a solution of G, too.

- Clearly the relations in $BOUND$ are satisfied.
- Clearly $\sigma(y) = \sigma(\Delta y)$ for all $y \in V$.
- The base equations are satisfied since $\sigma(x) = \sigma(x_0)$.
- Since $\sigma(b_0)s_i = \sigma(b_{i1}), \sigma(\beta_l(\Delta x_0))s_i = \sigma(b_i^*), \sigma(\beta_l(\Delta x))s_i = \sigma(b_{i2})$ σ satisfies $(b_{i1}/x/b_{i2})$ in G.
- Clearly all restrictions in $REST$ are satisfied.

9 Transformation of case C2

Let $x_0 \in V$ with $\beta_l(x_0) = b_0, \beta_l(\Delta x_0) \neq b_0$ and there exists no $b_0 x \asymp b_r \in BASE$ with $x \neq x_0$.
Let $b_I = \min\{b \in B : b > b_0\}$. Since each boundary must be right boundary or occur in a boundary connection there are two possible subcases

1. $b_I = \beta_r(x_0)$
2. there is a boundary connection $(b_I/x_0/b_{II}) \in CONN$.

If $b_I = \beta_r(x_0)$ then let $b_{II} = \beta_r(\Delta x_0)$ and define the base equations $e_0 = b_I\epsilon \asymp b_I, e_\Delta = b_{II}\epsilon \asymp b_{II}$ otherwise define $e_0 = b_I x_0 \asymp \beta_r(x_0), e_\Delta = b_{II}\Delta x_0 \asymp \beta_r(\Delta x_0)$.
The transformation consists of the following:

> IF $CONN$ contains boundary connections $(b_1/x_0/b_2), (b_1/x_0/b_3)$ with $b_2 \neq b_3$
> THEN RETURN (FALSE).
> IF $\exists(b_1/x_0/b_2) \in CONN : b_1 > b_I, b_2 \leq b_{II}$
> THEN RETURN (FALSE).
> OTHERWISE

$$
\begin{aligned}
BASE_T &:= BASE - \{b_0 x \asymp \beta_r(x_0), \beta_l(\Delta x_0)\Delta x_0 \asymp \beta_r(\Delta x_0)\} \\
&\quad +\{e_0, e_\Delta\} \\
CONN_T &:= CONN - \{(b_I/x_0/b_{II})\} \\
REST_T &:= REST - \{(b_0, \neg a, b_r) \in REST\} \\
&\quad +\{(\beta_l(\Delta x_0), \neg a, b_{II}) : (b_0, \neg a, b_r) \in REST\} \\
&\quad +\{(b_I, \neg a, b_r) : (b_0, \neg a, b_r) \in REST, b_I \neq b_r\} \\
BOUND_T &: \; B_T := B - \{b_0\} \\
VAR_T &:= VAR \;\; (\text{or } VAR - \{x_0, \Delta x_0\} \text{ if } b_I = \beta_r(x_0)).
\end{aligned}
$$

9.1 Proof for case C2

If there two boundary connections $(b_1/x_0/b_2), (b_2/x_0/b_3) \in CONN$ with $b_2 \neq b_3$ then G has no solution since $\sigma(b_0)s_2 = \sigma(b_1)$, $\sigma(\beta_l(\Delta x_0))s_2 = \sigma(b_2)$, $\sigma(b_0)s_3 = \sigma(b_1)$, $\sigma(\beta_l(\Delta x_0))s_3 = \sigma(b_3)$ implies $\sigma(b_2) = \sigma(b_3)$ which contradicts $b_2 < b_3$ or $b_2 > b_3$.
If there is a boundary connection $(b_1/x_0/b_2) \in CONN$ with $b_1 > b_I$ then a solution σ must satisfy $\sigma(b_0)st = \sigma(b_I)t = \sigma(b_1), \sigma(\beta_l(\Delta x_0))st = \sigma(b_{II})t = \sigma(b_2)$ which contradicts $b_2 \leq b_{II}$.
Clearly b_I is the minimal boundary of G_T and it is easily seen that G_T is a generalized equation and $f(G_T) \leq f(G)$.
We have to prove that b_{II} is right boundary or appears in a boundary connection in $CONN_T$:

> Clearly $(b_I/x_0/b_{II})$ is the only bondary connection in $CONN$ which contains b_I. If b_{II} is no right bounday then there must be boundary connections

con_0, con_r with $con_0 \trianglelefteq (b_I/x_0/b_{II})$ and $con_0 \trianglelefteq con_r$. Hence there must be a boundary connection $con \neq (b_I/x_0/b_{II})$ which contains b_{II} and PROP1 is satisfied.

9.1.1 G_T satisfies PROP2

Let $con \in CONN_T$. We have to prove that there are $con_0, con_r = (b_a/x/b_b) \in CONN_T$ with $con_0 \trianglelefteq_{G_T} con, con_0 \trianglelefteq_{G_T} con_r$ and $b_b = \beta_r^T(y), \beta_l^T(\Delta x) \leq \beta_l^T(y)$ or $b_b\epsilon \asymp b_b \in BASE$ or $con_0 = con_r$.
Since $con \in CONN$ we have $con_0, con_r = (b_a/x/b_b) \in CONN$ with $con_0 \trianglelefteq_G con, con_0 \trianglelefteq_G con_r$ and $b_b = \beta_r(y), \beta_l(\Delta x) \leq \beta_l(y)$ or $b_b\epsilon \asymp b_b \in BASE$ or $con_0 = con_r$. There is a chain

$$(b_1/x_1/b_2), (b_2/x_2/b_3), \ldots, (b_n/x_n/b_{n+1}))$$

where $con = (b_1/x_1/b_2)$, $con_r = (b_n/x_n/b_{n+1})$ and $con_0 = (b_k/x_k/b_{k+1})$ for some $k \in \{1, \ldots, n\}$. Furthermore

$$\begin{array}{ll} \beta_l(\Delta x_i) \geq \beta_l(x_{i+1}) & 1 \leq i \leq k-1 \\ \beta_l(\Delta x_i) \leq \beta_l(x_{i+1}) & k \leq i \leq n \quad (\beta_l(x_{n+1}) = b_b \text{ if } b_b\epsilon \asymp b_b \in BASE) \end{array}$$

for $k < n$ or

$$\beta_l(\Delta x_i) \geq \beta_l(x_{i+1}) \quad 1 \leq i \leq n-1$$

for $k = n$ (then $con_0 = con_r$).
Assume that in the chain there is an $i \in \{2, \ldots, n\}$ with $\Delta x_{i-1} = x_i$. Then a new chain can be obtained by deleting those boundary connections with x_{i-1}, x_i since

$$\begin{array}{l} \beta_l(\Delta x_{i-2}) \leq \beta_l(x_{i-1}), \beta_l(\Delta x_i) \leq \beta_l(x_{i+1}) \Rightarrow \beta_l(\Delta x_{i-2}) \leq \beta_l(x_{i+1}) \\ \beta_l(\Delta x_{i-2}) \geq \beta_l(x_{i-1}), \beta_l(\Delta x_i) \geq \beta_l(x_{i+1}) \Rightarrow \beta_l(\Delta x_{i-2}) \geq \beta_l(x_{i+1}). \end{array}$$

Hence we assume that there are no such pairs in the chain.
Assume that b_I appears in the chain. Since $(b_I/x_0/b_{II})$ is the only boundary connection which contains b_I there must be a pair $(b_{II}/\Delta x_0/b_I), (b_I/x_0/b_{II})$ in the chain which contradicts the above assumption.
Now assume that $x_0 = x_i$ or $\Delta x_0 = x_i$ appears in the chain:

Lemma 7 *If $x_i = x_0$ for some $i \in \{2, \ldots, n+1\}$ then $\beta_l(\Delta x_{i-1}) > \beta_l(x_i)$ and $\beta_l^T(\Delta x_{i-1}) \geq \beta_l^T(x_i)$.*

Proof. If $\beta_l(\Delta x_{i-1}) \leq \beta_l(x_i) = \beta_l(x_0) = b_0$ then $x_{i-1} = \Delta x_0$. Clearly $\beta_l^T(\Delta x_{i-1}) \geq b_I = \beta_l^T(x_0)$.

□

Lemma 8 *If $x_i = \Delta x_0$ for some $i \in \{1, \ldots, n\}$ then $\beta_l(\Delta x_i) < \beta_l(x_{i+1})$ and $\beta_l^T(\Delta x_i) \leq \beta_l^T(x_{i+1})$.*

Proof. Analogously to lemma 7.

□

Hence we are finished since either

$$\beta_l(\Delta x_{i-1}) > \beta_l(x_0), \beta_l(\Delta x_0) \geq \beta_l(x_{i+1})$$
$$\beta_l^T(\Delta x_{i-1}) \geq \beta_l^T(x_0), \beta_l^T(\Delta x_0) \geq \beta_l^T(x_{i+1})$$

or

$$\beta_l(\Delta x_{i-1}) > \beta_l(x_0), \beta_l(\Delta x_0) \leq \beta_l(x_{i+1})$$
$$\beta_l^T(\Delta x_{i-1}) \geq \beta_l^T(x_0), \beta_l^T(\Delta x_0) \geq / \leq \beta_l^T(x_{i+1}).$$

Analogously we treat $x_i = \Delta x_0$.

9.1.2 Let σ be a short solution of G

Clearly $\sigma(b_0)\sigma(x_0) = \sigma(b_I)s_x$ for some s_x where $s_x = \epsilon$ if $b_I = \beta_r(x_0)$. We construct a solution $\hat{\sigma}$ of G_T:

$$\hat{\sigma}(x) = \sigma(x) \quad \forall x \in V - \{x_0, \Delta x_0\}$$
$$\hat{\sigma}(b) = \sigma(b) \quad \forall b \in B$$
$$\hat{\sigma}(x_0) = \hat{\sigma}(\Delta x_0) = s_x \quad (\text{if } s_x \neq \epsilon).$$

- Clearly $\hat{\sigma}$ satisfies all conditions of $BOUND_T$.
- Clearly $\hat{\sigma}(y) = \hat{\sigma}(\Delta y)$ for all $y \in V_T$.
- $\hat{\sigma}(b_I x_0) = \sigma(b_0 x_0) = \sigma(\beta_r(x_0)) = \hat{\sigma}(\beta_r(x_0))$. Since $\sigma(b_0)s_1 = \sigma(b_I)$ and $\sigma(\beta_l(\Delta x_0))s_1 = \sigma(b_{II})$ implies $\sigma(x_0) = s_1 s_x$ we have $\hat{\sigma}(b_{II}\Delta x_0) = \sigma(b_{II})s_x = \sigma(\beta_l(\Delta x_0))s_1 s_x = \sigma(\beta_l(\Delta x_0))\sigma(x_0) = \sigma(\beta_r(\Delta x_0)) = \hat{\sigma}(\beta_r(\Delta x_0))$.
- Let $(b_1/x_0/b_2) \in CONN_T$. Then $(b_1/x_0/b_2) \in CONN$ and $b_1 > b_I$ and $\sigma(b_0)s_1 t = \sigma(b_I)t = \sigma(b_1), \sigma(\beta_l(\Delta x_0))s_1 t = \sigma(b_{II})t = \sigma(b_2)$. Hence $\hat{\sigma}(b_I)t = \hat{\sigma}(b_1), \hat{\sigma}(b_{II})t = \hat{\sigma}(b_2)$.
- Since $\sigma(b_0)s_1 t = \sigma(b_I)t = \sigma(b_r)$ and $\sigma(\beta_l(\Delta x_0))s_1 = \sigma(b_{II})$ all restrictions in $REST_T$ are satisfied.
- Clearly R#$(\hat{\sigma}, V_T, REP)$ holds.

Since $b_0 < b_I$ clearly $g_0(G_T, \hat{\sigma}) < g_0(G, \sigma)$.

9.1.3 Let σ be solution of G_T

Since $\beta_l(\Delta x_0) < b_{II}$ we have $\sigma(\beta_l(\Delta x_0))s_1 = \sigma(b_{II})$. Since $b_I < b$ for all $b \in B_T, b \neq b_I$ we have $\sigma(b) = \sigma(b_I)s_b$. We construct a solution $\hat{\sigma}$ of G:

$$\hat{\sigma}(x) = \sigma(x) \quad \forall x \in V - \{x_0, \Delta x_0\}$$
$$\hat{\sigma}(b) = \sigma(b_I)s_1 s_b \quad \forall b \in B - \{b_0, b_I\}$$
$$\hat{\sigma}(b_0) = \sigma(b_I)$$
$$\hat{\sigma}(b_I) = \sigma(b_I)s_1.$$

If $b_I < \beta_r(x_0)$ then

$$\hat{\sigma}(x_0) = \hat{\sigma}(\Delta x_0) = s_1\sigma(x_0)$$

else

$$\hat{\sigma}(x_0) = \hat{\sigma}(\Delta x_0) = s_1.$$

- Clearly $\hat{\sigma}$ satisfies all conditions of $BOUND$.
- Clearly $\hat{\sigma}(y) = \hat{\sigma}(\Delta y)$ for all $y \in V$.
- $\hat{\sigma}(b_0 x_0) = \sigma(b_I)s_1\sigma(x_0) = \hat{\sigma}(\beta_r(x_0))$ since $\sigma(\beta_r(x_0)) = \sigma(b_I)\sigma(x_0)$.
  $\hat{\sigma}(\beta_l(\Delta x_0)\Delta x_0) = \sigma(b_I)s_1 s_{\beta_l(\Delta x_0)} s_1 \sigma(x_0) = \hat{\sigma}(\beta_r(\Delta x_0))$ since $s_{\beta_l(\Delta x_0)} s_1 = s_{b_{II}}$ and $s_{b_{II}}\sigma(x_0) = s_{\beta_r(\Delta x_0)}$.
- $\hat{\sigma}(b_0)s_1 = \hat{\sigma}(b_I)$, $\hat{\sigma}(\beta_l(\Delta x_0))s_1 = \sigma(b_I)s_1 s_{\beta_l(\Delta x_0)} s_1 = \sigma(b_I)s_1 s_{b_{II}} = \hat{\sigma}(b_{II})$.
  Let $(b_1/x_0/b_2) \in CONN$, $b_1 \neq b_I$. Then $(b_1/x_0/b_2) \in CONN_T$ and $\sigma(b_I)t = \sigma(b_1)$, $\sigma(b_{II})t = \sigma(b_2)$. Hence $\hat{\sigma}(b_0)s_1 t = \sigma(b_I)s_1 t = \hat{\sigma}(b_1)$ and $\hat{\sigma}(\beta_l(\Delta x_0))s_1 t = \sigma(b_I)s_1 s_{\beta_l(\Delta x_0)} s_1 t = \sigma(b_I)s_1 s_{b_{II}} t = \hat{\sigma}(b_2)$.
- Since $\sigma(\beta_l(\Delta x_0))s_1 = \sigma(b_{II})$ and $\sigma(b_I)t = \sigma(b_r)$ we have $\hat{\sigma}(b_0)s_1 t = \sigma(b_I)s_1 t = \hat{\sigma}(b_r)$ and all restrictions in $REST$ are satisfied.

10 Transformation of case C3

Let for all $x_0 \in V$ with $\beta_l(x_0) = b_0, \beta_l(\Delta x_0) \neq b_0$ there be an $b_0 x \asymp b_r \in BASE$ with $x \neq x_0, b_r > \beta_r(x_0)$ or $x = \epsilon$. Furthermore let $b_I = \min\{b \in B : b > b_0\}$. We distinguish four subcases:

1. $\exists b_0 a \asymp b_r \in BASE : a \in C, b_r > b_I$.
2. $\exists b_0 \epsilon \asymp b_0 \in BASE$.
3. $\exists x_0 \in V : \beta_l(x_0) = \beta_l(\Delta x_0) = b_0$.
4. $\nexists x_0 \in V : \beta_l(x_0) = b_0$.

Clearly there are no more possibilities in case C3.

10.1 Subcase 1: $\exists b_0 a \asymp b_r \in BASE : a \in C, b_r > b_I$

The transformation returns FALSE.
A solution must satisfy $\sigma(b_0 a) = \sigma(b_0)a = \sigma(b_r)$ and $\sigma(b_0)st = \sigma(b_I)t = \sigma(b_r)$. This is not possible since $|st| > |a|$.

10.2 Subcase 2: $\exists b_0\epsilon \asymp b_0 \in BASE$

The transformation consists of the following:

$$\begin{aligned} BASE_T &:= BASE - \{b_0\epsilon \asymp b_0\} \\ BOUND_T &:= BOUND \\ VAR_T &:= VAR \\ CONN_T &:= CONN \\ REST_T &:= REST \end{aligned}$$

Clearly σ solves G if and only if it solves G_T, and $g_0(G_T, \sigma) < g_0(G, \sigma)$.

10.3 Subcase 3: $\exists x_0 \in V : \beta_l(x_0) = \beta_l(\Delta x_0) = b_0$

The transformation consists of the following:

<u>IF</u> $\beta_r(x_0) \neq \beta_r(\Delta x_0)$ <u>THEN</u> <u>RETURN</u> (FALSE).
<u>IF</u> $\exists(b_1/x_0/b_2) \in CONN : b_1 \neq b_2$ <u>THEN</u> <u>RETURN</u> (FALSE).
<u>OTHERWISE</u>

$$\begin{aligned} BASE_T &:= BASE - \{b_0x_0 \asymp \beta_r(x_0), b_0\Delta x_0 \asymp \beta_r(x_0)\} \\ &\quad +\{\beta_r(x_0)\epsilon \asymp \beta_r(x_0)\} \\ CONN_T &:= CONN - \{(b_1/x/b_1) : x = x_0\} \\ REST_T &:= REST \\ BOUND_T &:= BOUND \\ VAR_T &:= VAR - \{x_0, \Delta x_0\} \end{aligned}$$

10.4 Proof for subcase 3

Scanning the definitions it is easily seen that G_T is a generalized equation and that $f(G_T) \leq f(G)$.
If there was a boundary connection $(b_1/x_0/b_1) \in CONN$ and b_1 is no right boundary then there must be $con_0, con_r \in CONN$ with $con_0 \trianglelefteq (b_1/x_0/b_1), con_0 \trianglelefteq con_r$. Hence there must be other boundary connections with b_1, too, and G_T satisfies PROP1.
Since $con_1 \preceq (b_1/x/b_1) \preceq con_2$ implies $con_1 \preceq con_2$ boundary connections $(b_1/x/b_1)$ can be removed without violating PROP2.
If $\beta_r(x_0) \neq \beta_r(\Delta x_0)$ then a solution σ must satisfy $\sigma(b_0x_0) = \sigma(\beta_r(x_0)), \sigma(b_0x_0) = \sigma(\beta_r(\Delta x_0)), \sigma(\beta_r(x_0)) \neq \sigma(\beta_r(\Delta x_0))$ which is not possible.
If $(b_1/x_0/b_2) \in CONN$ with $b_1 \neq b_2$ then $\sigma(b_0)s = \sigma(b_1)$ and $\sigma(b_0)s = \sigma(b_2)$ which contradicts $\sigma(b_1) \neq \sigma(b_2)$.
If σ is a short solution of G then trivially σ is also a short solution of G_T. Since $b_0 < \beta_r(x_0)$ clearly $g_0(G_T, \sigma) < g_0(G, \sigma)$.

10.4.1 Let σ be solution of G_T

Since $b_0 < \beta_r(x_0)$ we have $\sigma(b_0)s_x = \sigma(\beta_r(x_0))$. We construct a solution $\hat{\sigma}$ of G:

$$\hat{\sigma}(x) = \sigma(x) \quad \forall x \in V - \{x_0, \Delta x_0\}$$
$$\hat{\sigma}(x_0) = \hat{\sigma}(\Delta x_0) = s_x$$
$$\hat{\sigma}(b) = \sigma(b) \quad \forall b \in B$$

- Clearly $\hat{\sigma}$ satisfies all conditions of $BOUND$.
- Clearly $\hat{\sigma}(y) = \hat{\sigma}(\Delta y)$ for all $y \in V$.
- $\hat{\sigma}(b_0 x_0) = \sigma(b_0)s_x = \sigma(\beta_r(x_0)) = \hat{\sigma}(\beta_r(x_0))$.
- The boundary connections $(b_1/x_0/b_1)$ are trivially satisfied.
- Clearly the restrictions in $REST$ are satisfied.

10.5 Subcase 4: $\nexists x_0 \in V : \beta_l(x_0) = b_0$

The transformation consists of the following:

<u>IF</u> $\exists b_0 a_1 \asymp b_1, b_0 a_2 \asymp b_2 \in BASE : a_1, a_2 \in C, a_1 \neq a_2$
<u>THEN</u> <u>RETURN</u> (FALSE).
<u>IF</u> $\exists b_0 a \asymp b_1 \in BASE : a \in C, b_1 \neq b_I$
<u>THEN</u> <u>RETURN</u> (FALSE).
<u>IF</u> $\exists b_0 a \asymp b_1 \in BASE, (b_0, \neg a, b_2) \in REST$ <u>THEN</u> <u>RETURN</u> (FALSE).
<u>OTHERWISE</u>

$$\begin{aligned}
BASE_T &:= BASE - \{b_0 a \asymp b_1 : a \in C \cup \{\epsilon\}\} - \{b_I \epsilon \asymp b_I\} \\
REST_T &:= (\; REST - \{(b_0, \neg a, b_1) : b_1 > b_0\} \\
&\qquad + \{(b_I, \neg a, b_1) : (b_0, \neg a, b_1) \in REST, b_1 > b_I\}) \\
&\quad - \{(b_1, \neg a, b_2) : a \notin \mathrm{const}(BASE_T)\} \\
BOUND_T &: B_T := B - \{b_0\} \\
VAR_T &:= VAR \\
CONN_T &:= CONN
\end{aligned}$$

10.6 Proof for subcase 4

Scanning the definitions it is easily seen that G_T is a generalized equation which satisfies PROP1, PROP2 and that $f(G_T) \leq f(G)$.
If $b_0 a_1 \asymp b_1, b_0 a_2 \asymp b_2 \in BASE$ and $a_1 \neq a_2$, $a_1, a_2 \in C$ a solution σ must satisfy $\sigma(b_0)a_1 = \sigma(b_1), \sigma(b_0)a_2 = \sigma(b_2)$ which is not possible.
If $b_0 a \asymp b_1 \in BASE, a \in C, b_1 \neq b_I$ then $b_1 > b_I$ and $\sigma(b_0)st = \sigma(b_I)t = \sigma(b_1)$ which contradicts $\sigma(b_0)a = \sigma(b_1)$.
If $b_0 a \asymp b_1 \in BASE, (b_0, \neg a, b_2) \in REST$ then $\sigma(b_0)s = \sigma(b_2), a \not\leq s$ which contradicts $\sigma(b_0)a = \sigma(b_1)$.

10.6.1 Let σ be a short solution of G

We show that σ satisfies the restrictions $(b_I, \neg a, b_1)$ with $(b_0, \neg a, b_1) \in REST$, $b_1 > b_I$. This is true since $\sigma(b_0)st = \sigma(b_I)t = \sigma(b_1)$, $a \not\leq st$. So σ is a short solution of G_T, too. Since there must be a boundary equation with right boundary b_I we have $|BASE_T| < |BASE|$. Hence $g_0(G_T, \sigma) < g_0(G, \sigma)$.

10.6.2 Let σ be solution of G_T

If there is a boundary equation $b_0 a \asymp b_I \in BASE$ then fix a, else let a be a new constant not appearing in G. For all $b > b_I$ there are $s_b \in S$ with $\sigma(b_I)s_b = \sigma(b)$. We construct a solution $\hat{\sigma}$ of G:

$$\begin{aligned}
\hat{\sigma}(x) &= \sigma(x) \quad \forall x \in V \\
\hat{\sigma}(b) &= \sigma(b_I)as_b \quad \forall b \in B - \{b_0, b_I\} \\
\hat{\sigma}(b_I) &= \sigma(b_I)a \\
\hat{\sigma}(b_0) &= \sigma(b_I).
\end{aligned}$$

- Clearly $\hat{\sigma}$ satisfies all conditions of $BOUND$.
- Clearly $\hat{\sigma}(y) = \hat{\sigma}(\Delta y)$ for all $y \in V$.
- Let $b_0 a \asymp b_1 \in BASE$. If $a \in C$ then $b_1 = b_I$ and $\hat{\sigma}(b_0 a) = \sigma(b_I)a = \hat{\sigma}(b_I)$. If $a = \epsilon$ then $b_1 = b_0$ and $b_0\epsilon \asymp b_0$ is trivially satisfied.
- Clearly the boundary connections in $CONN$ are satisfied.
- Let $(b_1, \neg c, b_2) \in REST$. If $b_1 = b_0, b_2 > b_I$ then $\hat{\sigma}(b_1)as_{b_2} = \hat{\sigma}(b_I)s_{b_2} = \hat{\sigma}(b_2)$ and $c \neq a$, $(b_I, \neg c, b_2) \in REST_T$ imply $c \not\leq as_{b_2}$. If $b_1 = b_0, b_2 = b_I$ then the restriction is satisfied since $\hat{\sigma}(b_1)a = \hat{\sigma}(b_I)$ and $c \neq a$. If $b_1 > b_0$ then $\sigma(b_1)t = \sigma(b_2)$, $c \not\leq t$ and hence $\hat{\sigma}(b_1)t = \sigma(b_I)as_{b_1}t = \sigma(b_I)as_{b_2} = \hat{\sigma}(b_2)$.

11 Conclusion

Putting the obtained pieces together we have the algorithm of theorem 1. The main steps to solve a set of string equations P with restrictions R are the following:

1. From (P, R) construct a generalized equation G as in section 3.
2. From G construct the set $\mathcal{G}$ of totally proper generalized euqations as described in section 6.
3. Apply the decision algorithm of section 4 with the initial step

   $P := \emptyset;\ O := \mathcal{G};$

Final remark: We hope that the stepwise construction of the algorithm gives the reader a better understanding of main ideas of the algorithm.

References

[1] Peter Auer. Unification in the combination of disjoint theories. 1991. To be published in the proceedings of the IWWERT'91.

[2] Peter Auer. *Unification with associative functions.* PhD thesis, Technical University Vienna, 1992.

[3] Franz Baader and Klaus Schulz. Unification in the union of disjoint equational theories: combining decision procedures. 1991. To be published in the proceedings of the IWWERT'91.

[4] Joxan Jaffer. Minimal and complete word unification. *Journal of the Association for Computing Machinery*, 37(1):47–85, January 1990.

[5] G. S. Makanin. The problem of solvability of equations in a free semiproup. *Math. USSR Sbornik*, 32(2):129–198, 1977. English translation.

[6] G. S. Makanin. Recognition of the rank of equations in a free semigroup. *Math. USSR Izvestija*, 14(3):499–545, 1980. English translation.

[7] J. P. Pecuchet. *Equations avec constantes et algorithme de Makanin.* PhD thesis, Laboratoire d' informatique, Rouen, 1981.

[8] Klaus U. Schulz. *Makanin's Algorithm — Two Improvements and a Generalization.* Technical Report 91-39, CIS—Universität München, 1991.

LOP: Toward a new implementation of Makanin's Algorithm

H. Abdulrab

LITP, Institut Blaise Pascal, France

and

LMI/INSA de Rouen, BP 08, 76131 Mont Saint Aignan Cedex

E.m.: abdulrab@geocub.greco-prog.fr

Tel: 35-52-83-41

Fax: 35-52-84-10

Abstract:

In this paper we describe a new approach to use in a further new implementation of Makanin's Algorithm.

We first explain the programming paradigms that express in a very natural way all the programming concepts needed in such an implementation.

Then we describe a programming system LOP (Logic, Objects and Parallelism) aimed at integrating these paradigms.

1. Introduction and Motivations

We are interested here in a new approach to use in a further new implementation of Makanin's algorithm, which is completely different from the first one given in [Abdulrab, 90a].

First, what is wrong with the first one?

In fact, its approach is very conservative. It proceeds in the following way:

Consider a good programming language, with a good programming environment (for example: LISP), and implement the algorithm in this programming language.

The new approach is radically different. It proceeds as follows:

First find (perhaps invent) a programming language that expresses in a natural way ALL the programming concepts of Makanin's Algorithm, and implement the algorithm in this language. Such an implementation should be very clear and concise. Efficiency is not the ultimate goal here.

Let us now explain what the programming paradigms are, that this language must contain.

The main paradigms are:

1) Unification and declarative style

Unification is used to compute the solutions of word equations from the decidability procedure of Makanin's algorithm [Abdulrab, 87], as well as in several steps matching variables of word equations to constants. This proposes that logic programming should be the basic background of such an implementation.

2) Object-Oriented Programming

The main concept used in Makanin's Algorithm is the concept of *position equation.* It is an an abstract and complex object. Moreover, there are several types of position equations.

So, it is very interesting to use object-oriented paradigms within the framework of declarative style to define position equations.

This could be realized by defining the *class* of position equations in such a way:

defclass(Position_Equation, Variables, Constants, Boundaries, Connections, Duality_Relation, ...).

And by creating an *instance* X of this class by using:

make-instance(X, Position_Equation, ...).

This can allow to the operations on position equations, used by Makanin's algorithm, to be implemented in a very convenient way, by using declarative style and the possibility to pass a position equation as a unique parameter rather than using a long list of parameters corresponding to the long list of the components of position equation.

In addition, unification of two *instances* should be a built-in primitive. This will allow to describe the important test, used in Makanin's algorithm, of the *equivalence* of two position equations, by testing only if they are unified. So deciding if two position equations PE1 and PE2, *instances* of Position-Equation class, are *equivalent* can be done directly (i.e. PE1 = PE1 ⇔ PE1 is *equivalent* to PE2).

3) Parallelism

Makanin's algorithm contains a lot of parallelism. It is a very good example of a massively parallel algorithm. Almost all its computations could be realized in parallel: The computation of applicable schemes at the first step of the algorithm, the test of the equivalence of two position equations, testing if a position equation exists already in the tree of the algorithm, the transformations of position equations, etc.

So, the language should express concurrent computations in a natural manner, independently of whether it is implemented in a parallel machine or not.

4) Constraints programming

Position equation can be seen as a set of several *constraints*. The transformations of such a position equation is nothing other than adding new constraints to already existing constraints.

Fortunately, all these paradigms are well-known, and well studied. But, unfortunately, they exist often separately, in distinct programming systems.

So, a fundamental question is the following: How to integrate them in a unique programming system? In other wards, what is a good integration policy?

Several important mechanisms used in different areas of logic programming (sequential, concurrent, object-oriented, *etc.*) share a lot of common factors, such as unification, declarative style, *etc.* They can make up an evolutive integrated programming system, satisfying the following requirements:

1) Every new mechanism enhances those already available, inherits the maximum of their behaviors and preserves completely the state previous to the introduction of the mechanism. This allows a program to be run in a minimal set of mechanisms, thus avoiding the situation where all the programming system mechanisms, including the superfluous ones, are necessarily brought into play.

2) Common concepts of many languages can be extracted within a generic system. They can be distinguished by a particular instantiation of the common factor of their differences.

LOP is a programming system designed to satisfy these points. Its main purpose is to integrate some important mechanisms in sequential, concurrent, object-oriented and constraint logic programming. LOP1, presented in this paper, is restricted to a base of some fundamental mechanisms in the first two areas. The final goal of LOP is to present a homogeneous, extensible and very expressive system for research and teaching.

Object-Oriented Programming Paradigms: It is natural that object-oriented programming paradigms appear to be a LOP implementation tool. Basic principles

[Ferber, 91] of this methodology, such as *abstraction, construction by refinement*, and *local thinking* (as opposed to the *global control* of classical programming) satisfy the points 1) and 2) above. In addition, other available object-oriented tools such as *auxiliary methods* allow an elegant introduction of the programming environment within the core of the programming system. Unlike other systems, such an introduction can be added completely *outside* the code of the programming system core.

LOP1, presented here, consists of some hierarchical systems associated with available mechanisms. Each system is defined by its *classes*, such as *Clause, Literal, Resolvant, etc.*, and *methods* over these classes. In each new system, many classes and methods are inherited unchanged from ancestor systems. Other classes are simply subclasses having possibly a new *instance variable* for the sake of the added system, or some *:before, :around,* or *:after auxiliary methods* for the additional behavior devoted to the new system, or some partially redefined methods corresponding to the new functionality to be introduced.

The progress of LOP systems starts with *Basic System*, containing basic classes and methods. It is followed by *Pure Prolog* System. Two new systems, called *Practical Prolog*, and *Classical Prolog* achieve a classical sequential Prolog. Several *delaying* mechanisms are then added in a new system called *Prolog with Delay*. The *Guard* mechanism is then introduced. A new system, whose goal is to define *Guarded Prolog* is then presented. Some guarded languages, such as FGHC, FCP(:), *etc.*, are then obtained in new LOP systems, by some modifications of GP functionality.

LOP is implemented in CLOS (Common Lisp Object System) [Steele, 90].

LOP Interpreters and Metainterpreters: *Classical Prolog* System is expressive enough to allow the description of following added systems by metainterpreters enhancing available ones, or by interpreters written in previous systems. This technique is used here as a formal and elegant specification of each added system's functionality, and the method of its implementation. Effective implementation is done in the second step, as outlined above, by object-oriented techniques adding a new level which takes advantage of the "history" of the programming system, without ever rewriting or changing it. The first self-description step provides an *executable* algorithm which helps greatly to clear and accelerate the second step.

2. Basic System

Among the most significant classes and instance variables of *Basic System*, denoted by $System_1$, can be found the following classes:

- *Predicate* defined by *name*, and *clauses*. (*i.e.* the list of all clauses whose head predicate is *name*).

- *Literal* defined by *term*.

- *Sequence-of-literals* defined by a list of objects of type *Literal*.

- *Clause* defined by *head* of type *Literal*, and *body* of type *Sequence-of-literals*.

- *Resolvant* defined by *literals* of type *Sequence-of-literals, applicable-clauses, remaining-clauses, goal, etc.*, where *applicable-clauses*, and *remaining-clauses* are those associated with the selected literal *goal.*

- *System* which describes the definition of a LOP system. Among its instance variables there are the names of classes composing the system, such as *class-of-literal, class-of-resolvant, etc.*, and *name, predefined-predicates, initial-goal, metainterpreter, interpreter, trace-p, etc.*, where *trace-p* is bound to *true* if the system is in *trace* mode, and to *false* otherwise.

Methods in System_1 define basic operations over these classes, such as adding new clause, creating and initializing resolvants, *etc.*, the *Solve* method in System_1 does nothing. It will be redefined in *Pure Prolog* System. An *auxiliary method* is associated with the *Solve* method of System_1 allowing instance variable values of resolvants to be displayed whenever a system is in *trace* mode.

$System_1$ is a subclass of *System* characterizing System_1. Every system of LOP is defined by a class $System_i$, subclass of $System_{i-1}$. Unique global variables of LOP are *Current-System*, and *System_1*, System_2*, *etc.*, instances of classes $System_1$, $System_2$, *etc.* Whenever *Current-System* is bound to *System_i*, only the mechanisms of System_1, ..., System_i, are available.

3. Pure Prolog

Pure Prolog System, denoted by System_2 contains the new subclass *Pure-Resolvant* of *Resolvant*, which has the new instance variable *current-substitution*. This choice of using *current-substitution* as *first class object*, is based on one of object-oriented programming principles, aiming to "abolish" global control of traditional programming methodology, by enhancing object representation and using a "local thinking" approach. Two new *internal* predicates are consequently defined:

- $environment(X): X is bound, by this call, to *current-substitution* of the current resolvant.

- $set-environment(X): *current-substitution* of the current resolvant is assigned, by this call, to X.

These predicates should be used with precaution. They could be called, for example, when a programmer decides to abandon an attempt to use a clause and wants to remove (without copies of terms) all effects on logic variables resulting from this attempt. They provide a simple solution in this case, as shown later in Program 7.2.

The code of the *Solve* method, over the *Pure-Resolvant* class is shown, for example, in Program 3.1. One can observe that classical parameters such as *current-substitution, literals, etc.*, implying global control, have disappeared. The method *end* over *Pure-Resolvant* is an *empty* method (it does nothing). This makes it possible to redefine *end* over some *most specific* subclasses of *Pure-Resolvant* without changing the methods in which it is called (as shown in §7.4).

```
(defmethod (Solve Pure-Resolvant) ()
   (if (null? self) (return-solutions self) (try-clauses self)))
(defmethod (Try-clauses Pure-Resolvant) ()
   (cond ((null remaining-clauses) (end self))
         (t (select-new-clause self) (reduce-by-selected-clause self)(try-clauses self))))
```

Program 3.1: Solve method.

4. Practical Prolog

In order to make *Pure Prolog* System practical, predefined predicates (such as arithmetic operations, type tests, *etc.*) must be defined in an optimized way. A new LOP system, called *Practical Prolog* denoted by System_3 is defined. It has a subclass of *Pure-resolvant*, called *Practical-resolvant*, with no new instance variable. Two new *internal* predicates, called $set-variable and $system, are defined in System_3. They allow definitions of predefined predicates to be imported from the language used to implement LOP (here Common Lisp). The definition of these two predicates is given below.

$set-variable(X, Expression).

$system(Expression).

A specific non logical reduction is applied to any term of this form, as follows.

-Expression, in $set-variable(X, Expression), is interpreted as a call to a Lisp function, and evaluated by *eval*. Its value is then unified with X. Note that variables in Expression should be *ground* terms. Some predefined predicates can be introduced by $set-variable, such as *plus* defined by:

```
plus(X, Y, Z)      :-      $set-variable(Z, (+ X Y)).
```

-Expression, in $system(Expression), must satisfy the same condition as in $set-variable. *Success* is obtained when the result of the evaluation is not *nil*. This predicate allows the introduction of many meta-logical and extra-logical predefined predicates, such as *integer* defined below. Observe that notations such as $system((integerp X)) are simplified.

```
integer(X)      :-      $system(integerp X).
```

The implementation of these two predicates is done by redefining *reduce-by-selected-clause* over *Practical-resolvant*. This consists in testing whether the goal to be reduced is one of these two special forms, in which case the special reduction, defined above, is done. Otherwise a "pure" reduction is applied by calling *reduce-by-selected-clause* of System_1. This is done by CLOS *call-next-method*, calling the next most specific method. Program 4.1 shows how this can be coded. The methods $system-p and $set-variable-p test whether any goal is a call of one these two special forms.

```
(defmethod (Reduce-by-selected-clause Practical-Resolvant) ()
```

```
(if (or ($system-p goal) ($set-variable-p goal))
    (special-reduction self)
    (call-next-method)))
```

Program 4.1: Redefinition of *reduce-by-selected-clause.*

It is clearly advisable that these two predicates should not be used as a "window" through which some pieces of Lisp codes could enter to replace some Prolog programs, for otherwise a cacophonous mixture of different concepts would result.

5. Classical Prolog

Introducing the *cut* mechanism allows a classical sequential Prolog to be achieved. This is the purpose of a new system called *Classical Prolog*, denoted by $System_4$. Three new classes are added. The first one, *Classical-resolvant*, is a subclass of *Practical-resolvant* with no new instance variable. The second, *Clause-with-cut*, inherits from *Clause*. It has the new instance variable, *cut-p,* to bind to *true* or *false* according to whether a *cut* occurs or not, respectively. The last one, *literal-cut,* inherits from *literal.* Its instances are calls of *cut.* A simple redefinition of *Solve* on *Classical-resolvant* using Lisp *catch* and *throw* is given in the full paper [Abdulrab, 91].

6. Prolog with Delaying Mechanisms

6.1. Freeze and Delay

The freezing mechanism, introduced by PrologII [Colmerauer, 82] via the predicate freeze(X, G), which consists in suspending a goal G (in a *Freezer*) until the variable X is *instantiated*, is a very interesting extension to Prolog, providing several programming techniques and applications [Colmerauer, 82 ; 87]. Another predicate [Carlsson, 87], called delay(X, G), suspending the execution of G until X is *bound*, can be used. It allows *freeze* to be defined as follows.

```
freeze(X, P)  :-    var(X) , delay(X, freeze(X, P)).
freeze(X, P)  :-    nonvar(X), P.
```

It is important to observe that introducing these predicates may add *suspension* alternative to the classical *success* and *failure* alternatives. Unlike these two *stable* terminations, suspension is *unstable.* If more information is added, *success* or *failure* can be deduced. Note that *suspension* results whenever the empty goal is obtained, and there is at least one suspended goal in the *Freezer.*

Example 6.1: integer(N), defined in §3, can be generalized as follows:

```
integer(N)    :-    freeze(N, $system(integerp N)).
```

integer(7) succeeds, *integer([])* fails, and *integer(X),* where X is non-instantiated suspends. But, *integer(X), X = 7* succeeds and *integer(X), X = []* fails. ♦

Remark 6.1: Suspension of a goal G until some variables $X_1, X_2, \ldots, X_n$ are *all instantiated* or *bound* is represented by $P_1(X_1, P_2(X_2, \ldots, P_n(X_n, G)\ldots))$, where P_i is either *freeze* or *delay*. In this case, $P_2(X_2, \dots)$ is suspended until X_1 is *instantiated* (if P_1 is *freeze*), or *bound* if (P_1 is *delay*). ♦

6.2. Implementation

Adding all delaying mechanisms, given in section §6, is firstly described by a metainterpreter, detailed in the full paper [Abdulrab, 91], slightly different from Cohen's one [Cohen, 85]. It defines Solve(Goal, Freezer, NewFreezer) which solves Goal, where Freezer and NewFreezer contain suspended goals before and after solving Goal, respectively. This allows the definition of call(Goal, S), given in Program 6.1, where S is instantiated to *true* if Goal succeeds, or *suspension* if it suspends.

```
call(Goal, true)          :-   solve(Goal, [], NewFreezer), NewFreezer =[].
call(Goal, suspension)    :-   solve(Goal, [], NewFreezer), NewFreezer ≠[].
```

Program 6.1: call(Goal, S).

We describe now how, following our metainterpreter scheme, object-oriented implementation is done. Introducing *delay* (*freeze* is its immediate consequence) is the main goal of a new system, called *Prolog with Delay*, and denoted by $System_5$. A subclass *Resolvant-with-delay* of *Classical-resolvant* is defined in $System_5$. It has the new instance variable *Freezer* with the following form:

$$((X_1\ b_{11} \dots b_{1p}), \dots, (X_n\ b_{n1} \dots b_{nm}))$$

where X_i is a variable and $b_{i1} \dots b_{ik}$ are *all* suspended goals until X_i is *bound*. This form allows a conjunction of goals to be unfrozen simultaneously. Note that, according to Remark 6.1, any suspended goal has only one occurrence in *Freezer*.

delay is implemented in $System_5$ as shown by Program 6.2.

```
delay(X, G)     :-   nonvar(X) , ! , G.
delay(X, G)     :-   $system(add-delayed-goal X G).
```

Program 6.2: Implementation of delay.

(add-delayed-goal X G) proceeds by testing whether *Freezer* contains a sublist (X b_1 ... b_i). In this case G is added before b_1, otherwise (X G) is pushed to *Freezer*. Exploring *Freezer* in order to test whether there are some goals to unfreeze is realized as follows. The method *create-son*, defined on *Resolvant* of $System_1$ to create a new resolvant after the success of head unification, has an *:after* auxiliary method over *Resolvant-with-delay*, whose role is to visit *Freezer*, collecting *all* b_1 ... b_i of *all* (X b_1 ... b_i), when X is *bound*, and adding them at the top of *literals* of the new resolvant.

When an empty goal is found, the value of its *Freezer* is assigned to *Freezer* of *initial-goal* G_0, allowing a test of whether call(G_0, true) or call(G_0, suspension) is obtained.

6.3. delay/1 and freeze/1

delay(G) and freeze(G) are introduced in LOP to suspend a goal G until *at least* one of the variables of G is *bound* or *instantiated*. They will be very useful in next LOP concurrent systems. If G is a ground term, delay(G) and freeze(G) are solved identically as G. Otherwise, delay(G) (freeze(G), respectively) is suspended until at least one of the variables of G becomes *bound* (*instantiated*, respectively). G is then solved. They can be written in LOP as follows:

delay(G) :- variables-of(G, X), delay+(X, G).

freeze(G) :- variables-of(G, X), freeze+(X, G).

variables-of(G, X) instantiates X to the list of all variables of G. delay+(X, G) (freeze+(X, G), respectively) suspends G (unless X is empty), until at least one variable in X is *bound* (*instantiated*, respectively). *freeze+* can be deduced from *delay+* as done below, where all-variables+(X, true) succeeds if each element of the non-empty list X is a variable, and all-variables+(X, false) succeeds if X is empty or if it has a non-variable term.

freeze+(X, G) :- all-variables+(X, true), delay+(X, freeze+(X, G)).

freeze+(X, G) :- all-variables+(X, false), G.

delay+ is implemented in $System_5$ as shown by Program 6.3, where *add-to-freezer* is a function that pushes (X G) to *Freezer* of the current resolvant.

delay+(X, G) :- all-variables+(X, false) , ! , G.

delay+(X, G) :- $system(add-to-freezer X G).

Program 6.3: Implementation of delay+.

Note that there is no confusion between this new form ((X_1 ... X_n) G) of sublists of *Freezer* and the previous one (X_k G_1 ... G_m). Thus, each suspended goal has always exactly one occurrence in *Freezer*. In addition, the visit to *Freezer* by the *:after* auxiliary method, described above, associated with *create-son*, can easily distinguish between the two forms of sublists. Whenever the new form ((X_1 ... X_n) G) is found and at least one X_i is *bound*, G is added at the top of *literals* of the new resolvant.

7. Guarded Prolog

We introduce here a new concurrent guarded language. Guard mechanism is used as a unique Or-parallelism synchronization rule. Delaying mechanisms presented in §6 are used as a unique And-parallelism synchronization rule. It is very important to observe that delaying mechanisms of §6 play here an essential and systematic role, since conjunction of goals is solved here simultaneously (not by using *depth-first* search of previous sequential systems). The language is called Guarded Prolog (GP), to signify that it adds only guard mechanism to Prolog (with delaying mechanisms).

Flat Guarded Prolog (FGP) is a subset of GP in which all the predicates called in the guards belong to a predefined subset of predicates.

Operational Semantic: The transition system [Kliger et al., 88] describing the operational semantic of concurrent logic programs of flat guarded language can be used for FGP. Using this formalism, it is sufficient to describe when reduction of a goal G by a clause C = (A :- Guard | Body) produces *success*, *failure*, or *suspension*.

- If G ≠ p(X, L), where p ∈ {*freeze, delay, freeze+, delay+*}, then the reduction of G

a) succeeds, if G = A and Guard resolution succeeds.

b) suspends, if G = A and Guard resolution suspends. (for example, when Guard = integer(N), where N is non-instantiated).

c) fails, otherwise.

- If G = freeze(X, L) (delay(X, L), respectively), where X is *instantiated* (*bound*, respectively), then the reduction of G is equivalent to the reduction of L. Otherwise, it is suspended until X is *instantiated* (*bound*, respectively).

- If G = freeze+(X, L) (delay+(X, L), respectively), and there is at least one *instantiated* (*bound*, respectively) variable in the list X, then the reduction of G is equivalent to the reduction of L. Otherwise, it is suspended until at least one variable of X is *instantiated* (*bound*, respectively).

7.2. Comparison with other rules

Suppose that access to a variable X of a goal L_i (appearing in a conjunction of goals $L_1 \ldots L_n$) is restricted to input mode [Takeuchi and Furukawa, 86]. In almost all guarded languages reductions of L_i by all applicable clauses are attempted. In FGP there is no attempt until X is *instantiated* or *bound* by *other* L_j. The advantage of the FGP rule is obtained if there is no clause selected by those attempts and *commitment* is delayed until enough information is produced. Observe, in addition, that calls of delaying predicates of LOP must be used when X is to be *bound* or *instantiated* by *other* L_j. If the value of X comes from an attempt to use a clause C, delaying predicates are not to be called. Attempting to use C may suspend, as in other languages, until there is more information.

FGP synchronization rule is very close to that of FCP [Shapiro, 86]. Unlike FGHC [Ueda, 86 ; 90] for example, but like FCP, synchronization is associated with "data", not with "procedure" [Takeuchi and Furukawa, 86]. Many FCP programs can be rewritten in FGP by transforming a term T containing an occurrence of a read-only variable X? to the form freeze(X, T'), where T' is obtained from T by replacing all the occurrences of X? by X. Freeze annotation can be seen as a practical simplification of CP read-only annotation [Shapiro, 86] which, however, could offer a simple and elegant solution when it is used to protect data structures from outside [Kliger et al., 88].

The need for delay(X, L) is justified by some particular cases, such as the following clause, which unifies Z and X when X is identical to Y.

f(X, Y, Z) :- X == Y | X = Z.

Observe that *freeze(X, f(X, Y, 1)), X = Y* suspends. But *delay(X, f(X, Y, 1)), X = Y* succeeds by unifying X, Y and Z to 1.

7.3. LOP Implementation of FGP

We give in this section a LOP sequential interpreter of FGP. This could be seen as a formal specification of FGP and its implementation. Effective implementation is realized in §7.4, as usual, by object-oriented techniques.

7.3.1. Interpreter

Program 7.1 gives a LOP interpreter of FGP. As in [Shapiro, 83], processes are scheduled in a queue. solve-GP(G) calls system(G) to test if G is a call of a predefined predicate. In this case, call(G) solves G by previous LOP systems. Otherwise, it calls schedule(G, Head, Tail, NewHead, NewTail), where Head is instantiated to the queue and Tail is a variable to be instantiated with the next process to enqueue. NewHead and NewTail represent the queue after adding G. solve-GP(Head, Tail) dequeues a process from Head, tests whether it is a call of a predefined predicate or not. In the first case it is solved as before. In the second case, it is reduced by reduce(G, B) which instantiates Cs (by calling clauses(G, Cs)) to a copy of all applicable clauses, copies G in CG, and calls resolve(G, CG, Cs, S, B). S is a suspension indicator. It is instantiated to *suspension* if there is an applicable clause whose guard suspends. B is instantiated to the body of the committed clause.

The first clause of *resolve* deals with guard suspension of the first applicable clause C. Logical variables of CG could be instantiated by head unification and guard resolution of C. Consequently, a new copy CG' of G is done for the next applicable clause. The second clause of *resolve* indicates when there is a committed clause. G and its copy are then unified in order to instantiate B according to the variables of G. The third clause of *resolve* calls remaining applicable clauses when the first one fails. The last clause of *resolve* is related to the suspension of the goal G. It succeeds when there is no applicable clause, and when S is previously instantiated to *suspension*.

resolve-head-and-guard(G, H, Guard, S) unifies G and H, and solves Guard as a call of predefined predicates. When Guard resolution succeeds (suspends, respectively), S is instantiated, by call(Guard, S) defined in Program 6.2, to *true* (*suspension*, respectively).

```
solve-GP(G)        :-    system(G), !, call(G).
solve-GP(G)        :-    schedule(G, X, X, Head, Tail), solve-GP(Head, Tail).
solve-GP([], [])   :-    !.
solve-GP([G | Head], Tail)  :- system(G), !, call(G), solve-GP(Head, Tail).
solve-GP([G | Head], Tail)  :-
    reduce(G, B), schedule(B, Head, Tail, NewHead, NewTail), !,
    solve-GP(NewHead, NewTail).
schedule([], Head, Tail, Head, Tail)      :- !.
schedule([A | B], Head, Tail, Head2, Tail2) :- !,
    schedule(A, Head, Tail, Head1, Tail1),
    schedule(B, Head1, Tail1, Head2, Tail2).
schedule(A, Head, [A | Tail], Head, Tail).
reduce(G, B) :-
    clauses(G, Cs), copy(G, CG), resolve(G, CG, Cs, Suspension, B).
```

```
resolve(G, CG, [(H :- Guard | Body) | Cs], S, B)  :-
    resolve-head-and-guard(CG, H, Guard, suspension), copy(G, CG'),
    resolve(G, CG', Cs, suspension, B), !.
resolve(G, CG, [(H :- Guard | Body) | Cs], S, Body) :-
    resolve-head-and-guard(CG, H, Guard, true), G = CG, !.
resolve(G, CG, [ _ | Cs], S, Q)          :-      resolve(G, CG, Cs, S, Q).
resolve(G, CG, [], S, []) :-      S == suspension, delay(G).
resolve-head-and-guard(G, G, Guard, S)        :-      call(Guard, S).
call(Goal, true)            :-      solve(Goal, [], NewFreezer), NewFreezer = [].
call(Goal, suspension)      :-      solve(Goal, [], NewFreezer), NewFreezer ≠ [].
```

Program 7.1: LOP FGP Interpreter.

7.3.2. Using environment predicates

The clauses *reduce* and *resolve* of Program 7.1 can be redefined, using the two internal predicates $environment and $set-environment, described in §3, to avoid copies of terms. This redefinition can be done by removing all *copy*'s calls, all the occurrences of CG, and by rewriting the first clause of *resolve* as follows.

```
resolve(G, [(H :- Guard | Body) | Cs], S, B)     :-
    $environment(X), resolve-head-and-guard(G, H, Guard, suspension),
    $set-environment(X), resolve(G, Cs, suspension, B).
```

$environment(X) instantiates X to *current-substitution* before calling *resolve-head-and-guard*. $set-environment(X) allows X to be the *current-substitution* removing all possible effects on logical variables of G.

7.3.3. Associating Freezer with resolvants

Associating *Freezer* with resolvants, described in §6.2, rather than using it as a global parameter, has a very interesting consequence, shown in this subsection.

Suppose that *Freezer* is a global parameter, as in some Prolog with Freeze interpreters which are automatically called for solving unfrozen goals. If Program 7.1 is to run in such a Prolog with Freeze interpreter, one must ensure that solving unfrozen goals is done by solve-GP interpreter of Program 7.1. This could be done by defining freeze-GP in such a Prolog with Freeze interpreter, as follows:

```
freeze-GP(X, G)  :-      freeze(X, solve-GP(G)).
```

But this introduces here a particular problem. Unfrozen goals are solved separately, by the new calls of solve-GP(G). A new queue is used for each call. If such a new call loops infinitely, or has a non-bounded number of reductions, other goals could wait forever. This clearly contradicts And-Parallelism principle simulated here by a *breadth-first* search in the process tree. So, one should explicitly manipulate the freezer in Program 7.1, as described in the full paper, by adding two new parameters Freezer and NewFreezer to almost all the clauses, just as a freezer is used to enhance a classical Prolog's metainterpreter [Cohen, 85].

This important complication which avoids inheriting what is already defined is eliminated in LOP. All unfrozen goals are simply added to *literals* of the current resolvant, as described in §6.2. They are solved by the same interpreter as all other goals of the resolvant.

7.4. Object-Oriented Implementation of FGP

FGP is the goal of a new LOP system, denoted by $System_5$ and called Guarded Prolog. New classes in $System_5$ are:

- *GP-Resolvant*, a subclass of *Resolvant-with-delay*. It has the new instance variable *suspension-p* used as a suspension indicator.

- *GP-guard*, a subclass of *Resolvant-with-delay*. It has the new instance variable *resolvant* indicating the resolvant whose *goal* attempts to solve the guard.

- *GP-clause* a subclass of *Clause-with-cut*. It has the new instance variable *guard* of *GP-guard* type.

Three simple redefinitions of some previous methods are done to obtain GP.

1) The method *new-goals*, defined over *Pure-Resolvant*, to produce *literals*, by concatenating the body $B_1, \dots, B_n$ of the selected clause with remaining goals $R_2, \dots,$ R_m is redefined on *GP-Resolvant* to obtain the reversed concatenation $R_2, \dots, R_m, B_1,$ $\dots, B_n$. Note that there is no global queue parameter here. The instance variable *literals* of *Resolvant* plays this role.

2) The method *Reduce-by-selected-clause*, given in Program 3.1, is redefined to solve guards of applicable clauses. A new method *Solve-guard* is defined over *GP-guard* in order to add guard resolution as follows. It sends the message *Solve* to a guard G. If such a resolution succeeds, it sets *remaining-clauses* of the *resolvant* R associated with G to () in order to stop trying other applicable clauses, and it sets *suspension-p* of R to *false*. If the resolution of G suspends it sets *suspension-p* of R to *true*.

3) The *empty* method *end*, defined in Program 3.1 over *Pure-Resolvant* to do nothing, when all applicable clauses are attempted, is redefined over *GP-Resolvant* to test whether *suspension-p* is *true*. In this case *goal* is delayed by *delay/1*.

7.5. Implementation of other guarded languages

We show in this section how other guarded languages can be obtained by redefining *resolve-head-and-guard*. We use the same notations as FCP(:) [Kliger et al., 88], separating guard predicates to *Ask* and *Tell* predicates. *Ask* predicates can test values of variables but cannot instantiate them, whereas *Tell* predicates can instantiate them. Program 7.2 define solve-tell and solve-ask. solve-tell is identical to call(G, S) of Program 7.1. The first clause of solve-ask succeeds when head unification and guard resolution does not instantiate any variable of Goal to a non-variable term. all-variables*(Xs, true) succeeds if each element of the list Xs is a variable. all-variables*(Xs, false) is its negation. The second clause of solve-ask reports a suspension when head unification and guard resolution only succeed by instantiating some variables of the goal. The last clause of solve-ask reports a suspension when head unification succeeds and guard resolution suspends.

```
solve-tell(Guard, S)                          :-    call(Guard, S).
solve-ask(Goal, Head, Guard, true)            :-    variables-of(Goal, Xs),
    Goal = Head, solve-tell(Guard, true), all-variables*(Xs, true).
solve-ask(Goal, Head, Guard, suspension)      :-  variables-of(Goal, Xs),
    Goal = Head, solve-tell(Guard, true), all-variables*(Xs, false), !.
solve-ask(Goal, Goal, Guard, suspension)      :-  solve-tell(Guard, suspension).
```

Program 7.2: *Ask* and *Tell*.

Program 7.3 redefines *resolve-head-and-guard* to obtain a simple FGHC interpreter. Delaying predicates of FGP are not to be used in a "pure" FGHC version. Using them could add the advantage outlined in §7.2.

```
resolve-head-and-guard(Goal, Head, Guard, S) :-  solve-ask(Goal, Head, Guard, S).
```

Program 7.3: a simple FGHC interpreter.

Program 7.4 redefines *resolve-head-and-guard* to obtain a simple FCP(:) interpreter.

```
resolve-head-and-guard(Goal, Head, (Ask : Tell), true) :-
    solve-ask(Goal, Head, Ask, true), solve-tell(Tell, true).
resolve-head-and-guard(Goal, Head, (Ask : Tell), suspension) :-
    solve-ask(Goal, Head, Ask, suspension), !.
resolve-head-and-guard(Goal, Head, (Ask : Tell), suspension) :-
    solve-ask(Goal, Head, Ask, true), solve-tell(Tell, suspension).
```

Program 7.4: a simple FCP(:) interpreter.

Implementation of other guarded languages is given in the full paper.

7.6. Full GP interpreter

We finish this section by a short note on how to obtain a (full) GP interpreter (*i.e.* allowing user's predicate calls in the guard). First suppose that we have a Solve-GP(Goal, Freezer, NewFreezer) as discused in §7.3.3. It is then possible to redefine call(Goal, S), given in Program 7.1, in order to obtain a full GP interpreter. Solve-GP is called here rather than *Solve* of Prolog with Delay System, as follows.

```
call(Goal, true)          :-    solve-GP(Goal, [], NewFreezer), NewFreezer =[].
call(Goal, suspension)    :-    solve-GP(Goal, [], NewFreezer), NewFreezer ≠[].
```

FGP is then obtained when restriction to predefined predicates in the guards is imposed. Otherwise, a GP interpreter is deduced. Object-oriented implementation, described in §7.4, proceeds in the same manner. The method *Solve-guard* calls the most specific method *Solve*, *i.e.* that implementing GP. Restriction to flat guards is equivalent to the call of the next most specific method *Solve* .

solve and *schedule* clauses of Program 7.1 are based on the CP interpreter [Shapiro, 83] using a queue to schedule processes (every thing which concerns the CP synchronization rule is eliminated from these clauses). All other clauses implement the

GP synchronization rule. The CP interpreter does not use a Freezer for suspended goals. Suspended goals as well as failed goals are scheduled in the same queue as other goals. At every exploration of the queue, suspended goals are systematically solved (even if there is no new information). A failed goal is solved unnecessarily in the same manner, rather than halting the resolution whenever such a goal occurs. Using a freezer in our interpreter provides a very important optimization in the two cases. In addition, it allows, unlike the CP interpreter, a distinction between *failure* and *suspension*.

Note that, as in the CP interpreter, our sequential implementation of GP does not implement a correct parallel search for the committed clause. A guard with infinite loop, could block attempts to use other applicable clauses.

8. Conclusions and Comparison with Other Work

Since Makanin's algorithm uses all important types of numerical and symbolic computation, and the four important paradigms described at the beginning of this paper, we think that it should be used as a benchmark of the expressiveness of programming systems.

We exposed the first step of such a system: LOP, aimed at integrating these paradigms.

Our main goal here was to show that object-oriented programming is a very attractive and promising tool in the implementation and integration of several well-known mechanisms of logic programming. Some parts of this work have been compared in §7.2 and §7.6 with related work. Using a freezer for suspended goals is also studied in [Saraswat, 89 ; Shapiro, 89]. Our goal here was to use delaying mechanisms as a unique And-parallelism synchronization rule, to present a language based on this idea, to give its interpreter, and to deduce further flat guarded languages.

9. Further Research

At this stage of LOP development, efficiency is not the ultimate goal of our implementation. Efficiency of some parts could be improved further. In addition, we are interested in adding ideas from some extensions of concurrent languages such as Andorra [Haridi and Brand, 88]. Another important extension (partially achieved in LOP) is to introduce some object-oriented primitives of concurrent logic languages, in a similar way to Vulcan [Kahn et al., 87]. Introduction is done here at GP System, to be automatically inherited by other flat guarded languages of LOP. Our implementation approach is completely different from Vulcan's. Rather than writing a transducer to FGP, in order to obtain a Vulcan-like language, we essentially use the same code and the same representation of classes, instances, *etc.*, of our mini CLOS, in which LOP is implemented. (The code of some basic functions of this mini CLOS is given in [Abdulrab, 90b]). Using this object-oriented logic programming language to implement added LOP systems is our next goal.

Acknowledgements

thanks to J. Cohen, A. Colmerauer and C. Kirchner for their advice and encouragement.

REFERENCES

H. Abdulrab. "Résolution d'équations sur les mots: étude et implémentation LISP de l'algorithme de Makanin". (Thèse), Université de Rouen (1987). And Rapport LITP 87-25, Université de Paris-7, 1987.

H. Abdulrab. "Implementation of Makanin's algorithm". Proceedings of IWWERT'90, Tubingen, RFA, LNCS, Springer-Verlag N. 572, Edited by Prof. K. Schulz, 1990.

H. Abdulrab, "de Common Lisp à la Programmation Objet", Editions HERMES, 1990.

H. Abdulrab, "Logic, Objects and Parallelism", (full paper), Rapport Technique LITP 91/60, 1991.

M. Carlsson, "Freeze, Indexing, and Other Implementation Issues in the WAM", in *Logic Programming: Proceedings of the Fourth International Conference*, MIT Press, Vol. 1, pp. 40-58, 1987.

J. Cohen, "Describing Prolog by its interpretation and compilation", *Communications of ACM*, (12), Dec. 1985.

A. Colmerauer, "Prolog II: Manuel de Référence et Modèle Théorique", Groupe Intelligence Artificielle, Université Aix-Marseille II, 1982.

A. Colmerauer, "Une Introduction à PrologIII", GIA, Université Aix-Marseille II, 1987.

J. Ferber, "Conception et Programmation par Objets", Editions HERMES, 1991.

S. Haridi, and P. Brand, 1988. "Andorra Prolog: an integration of Prolog and committed choice languages". In *Proceedings of the International Conference on Fifth Generation Computer Systems*, pp.745-754, 1988.

K. M. Kahn and M. Carlsson "How to implement Prolog on a Lisp Machine", Implementations of Prolog, edited by J.A. Campbell, pp. 117-162, 1984.

K. Kahn, E.D. Tribble, M.S. Miller and D.G. Bobrow, "Vulcan: Logical Concurrent Objects", Chap. 30. in [Shapiro, 1987], 1987.

S. Kliger, E. Yardeni, K. Kahn and E. Shapiro, "The Language FCP(:,?)", Proceedings of the International Conference on Fifth Generation Computer Systems, pp. 763-773, 1988.

J.W. Lloyd, "Foundations of Logic Programming", Symbolic Computation series, Springer Verlag, Second Edition, 1987.

V. A. Saraswat, "Concurrent Constraint Programming Languages", Ph. D. Thesis, CMU, 1989.

E. Shapiro, "A subset of Concurrent Prolog and its interpreter", Technical Report, CS83-06, Weizmann Institute, 1983. (Also Chap. 2 in [Shapiro, 87]).

E. Shapiro, "Concurrent Prolog: a progress report", *IEEE Computer*, 19(8): 44-58, August 1986. (Also Chap. 5 in [Shapiro, 87]).

E. Shapiro, editor. "Concurrent Prolog", Volume 1 et 2. MIT Press, 1987.

E. Shapiro, " The family of concurrent logic programming language". Technical Report CS89-08, Departement of Applied Mathematics and Computer Science, The Wietzmann Institute, Rehovot. 1989.

G. L. Steele, "Common Lisp: the language", Digital-press, Second Edition, 1990.

A. Takeuchi, and K. Furukawa, "Parallel logic programming languages", In *Proceedings of the Third International Conference on Logic Programming*, pp. 242-254, July 1986.

K. Ueda, " Guarded Horn clauses", Ph. D. thesis, University of Tokyo, 1986.

K. Ueda, "Designing a Concurrent Programming Language", in Proceedings of an International Conference organized by the IPSJ to Commemorate the 30th Anniversary (InfoJapan'90), *Information Proceedings Society of Japan*, pp. 87-94, October 1990.

Word Unification and Transformation of Generalized Equations

Klaus U. Schulz

Centrum für Informations- und Sprachverarbeitung (CIS)
University Munich
Leopoldstr. 139, D-8000 Munich 40

Abstract. Makanin's algorithm [Ma77] shows that it is decidable whether a word equation has a solution. The original description was hard to understand and not designed for implementation. Since words represent a fundamental data type, various authors have given improved descriptions [Pé81, Ab87, Sc90, Ja90]. In this paper we present a version of the algorithm which probably cannot be further simplified without fundamentally new insights which exceed Makanin's original ideas. We give a transformation rule which is efficient, conceptually simple and applies to arbitrary generalized equations. No further subprocedure is needed for search tree generation. In contrast to our older work in [Sc90] the presentation will be based on Jaffar's [Ja90] notion of generalized equations. We also prove that a combination of Plotkin's algorithm (see [Pl72], also [Le72]) and Makanin's algorithm offers a simple solution to the problem of terminating minimal and complete word unification.

Introduction

One of the simpler tasks which a text editing system has to solve again and again is the problem to determine whether a particular string W occurs in a given text T. This problem may be expressed by means of an equation $T == xWy$. Obviously W occurs in T if and only if there exists words X and Y in the text alphabet which solve this equation, i.e., words X and Y such that T and XWY are identical. What we have to solve is a particular *word equation*. A word equation is an expression of the form $W_1 == W_2$, where $W_i \in (\mathcal{C} \cup \mathcal{V})^+$ are non-empty words in a mixed alphabet $\mathcal{C} \cup \mathcal{V}$ of constants and variables respectively. Of course the set of constants $\mathcal{C}$ and the set of variables $\mathcal{V}$ are disjoint. A *solution* of the word equation $W_1 == W_2$ is an assignment of values $X_i \in \mathcal{C}^+$ to the variables x_i occurring in the equation such that W_1 and W_2 become identical if all variables are replaced by the corresponding values. When the values X_i are allowed to be words in the mixed alphabet $(\mathcal{C} \cup \mathcal{X})^+$, then we get the notion of a *unifier* of the word equation.

The importance of the problem to decide whether a word equation has a solution/unifier becomes appearent if other formulations are used. Word equations may be called equations in a free semigroup, equations over lists (of atomic elements) with concatenation, or associative unification problems with constants, stressing their role in mathematical logic ([Hm71, Ma77], constraint logic programming ([Col88]) or universal unification theory ([Ba90, Si89]).[1]

Historically A.A. Markov, at the end of the 1950's, was probably the first to ask whether it is decidable if a word equation has a solution. Markov noted that every word equation over a two-letter constant alphabet may be translated into a finite system of diophantine equations, preserving solvability in both directions. He hoped to obtain a proof for the unsolvability of Hilbert's tenth problem by showing that solvability of word equations is an undecidable problem (see [Ma81] for more details). Approximately at the same time Lentin and Schützenberger [LeSc67] independently considered word equations.

In the following period, in the western countries the main attention was given to the (relatively simple) problem to enumerate in a compact form the set of all solutions of a word equation. Plotkin [Pl72], in the context of resolution based theorem proving, gave a simple algorithm to generate a minimal and complete set of unifiers (see section 5 for these notions) for a given word equation.[2] Lentin [Le72] independently found the same algorithm which was later rediscovered several times. In the eastern countries, the much harder decision problem was addressed. Hmelevskiĭ, [Hm71] obtained a partial solution, showing that the solvability of word equations with three variables is decidable. Later G.S. Makanin showed in his epochal paper [Ma77] that the solvability of arbitrary word equations is decidable.

The Problem of an Optimal Form. Makanin's algorithm is based on the new data type of *generalized equations*. It starts translating a word equation into a finite set of generalized equations. Then two subprocedures, "transformation" and "normalization," are used to generate a finitely branching search tree. In general, this subtree generation process does not stop, and Makanin uses an ingenious idea for termination.

The original decision procedure was not designed for implementation and is hard to understand. Several attempts have been made to find a better description [Pé81, Sc90, Ja90] and the algorithm has now been implemented [Ab87]. In this paper we shall present a version of Makanin's algorithm which probably cannot be further simplified without fundamentally new insights into the problem — insights which essentially exceed Makanin's original ideas. Our presentation will be based on J. Jaffar's [Ja90] modified notion of generalized equations which replaces Makanin's concept of a "boundary connection"

[1] As a matter of fact it is simple to decide solvability of *matching equations* like $T == xWy$, and efficient algorithms are known for these restricted problems (see, e.g., [Ah90]).

[2] Plotkin's algorithm was able to deal with a more general problem, namely unification of first order terms modulo associativity of a given function symbol.

by the concept of a "boundary equation." With this step, various important improvements are obtained: the normalization subprocedure which occurs in [Ma77, Pé81, Ab87, AbPe89, Sc90] is avoided and transformation of generalized equations becomes simpler since no boundary connections have to be updated. This source of complexity has been removed from the algorithm and is now situated in the correctness proof.

Jaffar's algorithm may be described as a tree generation process which is based on the iteration of two subprocedures, completion and transformation (= reduction). The former — trivial — algorithm transforms every generalized equation in an equivalent set of completed generalized equations. The latter procedure transforms a completed generalized equation into a simpler generalized equation. Jaffar, as well as [Ma77, Pé81, Ab87, AbPe89], distinguishes several types of completed generalized equations. For each type a special transformation rule is given.

In this paper we present an improved version of the transformation algorithm which has a built-in completion and consists of just one rule which applies to arbitrary completed generalized equations[3]. Thus, with the new transformation algorithm, the search tree generation process is based on one subprocedure only. From a conceptual point of view, the effect of a transformation step may be described very easily:

- At a transformation step, a *non-empty* left part of the generalized equation is *simultaneously* carried towards the right side of the generalized equation.

With this property the new transformation is very similar to the transformation steps which are used in Plotkin's (Lentin's) procedure. Our main aim, however, was to reduce generalized equations as efficient as possible, even for the price that proofs become more complex.

- In essence, one step of our algorithm combines all those steps of Jaffar's algorithm that are made with the same "carrier". Thus, a maximal number of bases and boundaries are transported simultaneously. In comparison with sequences of case dependent transformations a lot of redundant work is avoided. The permanent encoding and decoding of information in and from boundary equations is widely avoided and new boundary equations are only introduced in specific cases.

The simplicity of the new transformation makes it very natural and easy to implement.

---

[3] A similar procedure was introduced in [Sc90], based on Pécuchet's notion of a position equation. But it turns out that the procedure is much simpler if based on Jaffar's representation.

The Problem of a Complete and Correct Proof. The proof that Makanin's algorithm is correct and complete has its own history. It has turned out that the definition of a generalized equation has to contain two subtile conditions (see [Sc90], pg. 126) whose sense becomes only clear when technical details of the transformation algorithm are considered. Unfortunately, this point was not treated correctly in the "classical" papers on Makanin's algorithm. Already Makanin's description [Ma77] (at least in its English version) contained a misprint in the definition of a boundary connection[4]. Supported by the subsequent formulations in [Ma77], most readers were led to a real misinterpretation of this definition. In particular, Pècuchet's and Abdulrab's definition of a position equation in [Pé81, Ab87, AbPe89] follows such a misinterpretation and does not lead to a correct proof. Similar difficulties arise from Jaffar's [Ja90] notion of a proper generalized equation.

In the meantime, these points are wellknown among the experts in the field. But, to my knowledge, there is no journal publication which contains a complete and error-free proof showing that transformation of generalized equations behaves as it should behave. Such a proof will be given in the extended version [Sc92] of this paper where we show that proper generalized equations are transformed into proper generalized equations with our transformation rule.

The structure of the paper is as follows. In section 1 we shall first describe the basic algorithm for word unification which goes back to [Le72, Pl72]. The behaviour of this algorithm in a particular subcase will be helpful to understand one of the main ideas behind the definition of a generalized equation. This definition is given afterwards. In the third part of section 1 we will sketch how word equations are translated into sets of generalized equations. In section 2 we introduce some notation, and in section 3 we define the transformation algorithm and show that transformation preserves unifiability in both directions. We obtain a correct and complete procedure testing unifiability of word equations. In section 4 we discuss termination. For this purpose, the notion of a proper generalized equation is introduced. In section 5 we shall prove — following an idea of F. Baader — that a combination of Plotkin's and Makanin's algorithms gives a simple solution to the problem of *terminating minimal and complete word unification* which was first solved in [Ja90]. This is the problem to find an algorithm which generates a minimal and complete set of unifiers for a given word equation and terminates if this set is finite.

1 Word Equations and Generalized Equations

In this section we want to introduce the concept of a generalized equation and to show how word equations are translated into sets of generalized equations.

[4] In [Ma77], the last inequality of (3.11) should be $l_{\alpha(\lambda_k)} \geq l_{\alpha(\Delta(\lambda_k))}$. In later, less known papers [Ma80, Ma81], this misprint was eliminated but the notion of "convexity" given there was not quite correct.

As mentioned above we shall start with a description of the basic algorithm for word unification which was independently found by Lentin [Le72] and Plotkin [Pl72]. Later, in section 5, we shall prove that the algorithm computes a disjoint and complete set of unifiers, for every word equation WE_0. For the moment we are only interested in the general structure.

The Basic Algorithm

For given word equation WE_0, a finitely branching search tree $\mathcal{T}_{LP}(WE_0)$ is generated, using non-deterministic transformation rules based on "variable-splitting" techniques. With $Succ(WE)$ we will denote the set of successors of the word equation WE under transformation. If WE has the form $uW_1 == vW_2$ (where u and v are constants or variables and W_1, W_2 are possibly empty words), then we say that WE has the "head" (u, v) and the *tail* $Tail(WE) := W_1 == W_2$. Expressions $(v \mapsto W)$ denote substitutions which simultaneously replace all occurrences of v by W if applied to any word equation.

Transformation:

(i) If WE has head (a, b) with two distinct constants, then $Succ(WE)$ is empty.
(ii) If WE has head (u, u) with two identical constants or variables, then $Succ(WE) := \{Tail(WE)\}$.
(iii) If WE has head (a, x) or (x, a) with one constant and one variable, then let $S_1^{LP} = (x \mapsto a)$, $S_2^{LP} = (x \mapsto ax)$. Now $Succ(WE) := \{Tail(S_1^{LP}(WE)), Tail(S_2^{LP}(WE))\}$.
(iv) If WE has head (x, y) with two distinct variables, then let $S_3^{LP} = (y \mapsto x)$, $S_4^{LP} = (x \mapsto yx)$ and $S_5^{LP} = (y \mapsto xy)$. Now $Succ(WE) := \{Tail(S_3^{LP}(WE)), Tail(S_4^{LP}(WE)), Tail(S_5^{LP}(WE))\}$.

Successor elements may have one or two empty sides. Every node labelled with a "word equation" with two empty sides is a successful leaf. Every node labelled with a "word equation" with exactly one empty side is a blind leaf.

In this version the algorithm defines a semi-decision procedure: it is straightforward to see that WE_0 has a solution iff $\mathcal{T}_{LP}(WE_0)$ has a successful leaf (see section 5).

The Explosion of the Data Size. In general, the basic algorithm does not terminate since the size of the word equations which are created via transformation may grow. As an example, consider the word equation $xyxy == axyxzb$. We may apply the transformation $WE \rightarrow Tail(S_2^{LP}(WE))$. The resulting word equation $xyaxy == axyaxzb$ has again head (x, a). After k iterations we obtain the word equation $xya^kxy == axya^kxzb$. In some cases, termination arguments may be given based on splitting techniques (see [LiSi75]). In general, however, there is no simple additional technique to decide solvability by deciding when an infinite branch can be cut.

A Decidable Subcase. In a particular case it is simple to obtain a terminating algorithm by means of loop-checking methods: a trivial inspection of the transformation rules shows that the size of word equations cannot grow if no variable occurs more than twice. Thus, in this case there is only a finite number of word equations which may occur in the search tree. Suppose that we have reached — at node ν_2 — a word equation which has occurred earlier in the same path, at node ν_1. In this case we may stop with failure: if any sequence of transformations leads to a successful leaf, starting from ν_2, then we may apply the same sequence starting from ν_1 and we will again find a successful leaf. This shows that we may ignore the subtree below ν_2 for matters of decidability. More generally we may stop when we have found a word equation WE_2 which is *isomorphic* to a predecessor WE_1 in the same path. This means that WE_2 may be obtained from WE_1 by a permutation of the variable alphabet and a permutation of the alphabet of constants.

With this pruning method we obtain a *finite* subtree $\mathcal{T}_{LP}^{fin}(WE_0)$ and thus a decision procedure: WE_0 has a unifier iff $\mathcal{T}_{LP}^{fin}(WE_0)$ has a successful leaf.

Generalized Equations

The observation that unifiability of word equations is decidable if variables occur at most twice does not solve the general problem. We cannot translate an arbitrary word equation into an equivalent word equation where every variable has at most two occurrences. But, with a more complex data type, it is in fact possible to get an artificial duality of variables. This is one of the important ideas behind the following concept.

Definition 1.1. A *generalized equation* is a quadrupel $GE = (BS, BD, Col, BE)$ with four entities:

(1) A finite set of *bases*: $BS = \{bs_1, ..., bs_N\}$, $N \geq 1$.
Every base is either a variable base or a constant base. Each constant base bs_i is associated with exactly one letter a in the alphabet C, we say that bs_i has type a. Each variable base bs_i is paired off with exactly one other variable base $bs_j \in BS$; bs_i and bs_j are called *duals* of each other. Letters $x, y, z, \ldots$ denote variable bases. We write $\bar{x}$ for the dual of the variable base x.

(2) A finite set of *boundaries*: $BD = \{1, 2, \ldots, M\}$, $M \geq 1$.
Letters $i, j, k, \ldots$ denote boundaries. A pair (i, j) of boundaries with $i \leq j$ is called a *column* of *GE*. Columns (i, i) are called empty, columns $(i, i+1)$ are called indecomposable. For $i < j < k$ we say that boundary j is in (i, k).

(3) A column-function *Col*:
A function which assigns a column of *GE* to every base of *BS* such that $Col(bs_i)$ is indecomposable for every constant base $bs_i \in BS$ and that $Col(x) \neq Col(\bar{x})$ if x is a non-empty variable base[5]. For convenience, we introduce the two functions *Left* and *Right*: if $Col(bs_i) = (j, k)$, then $Left(bs_i) = j$ and $Right(bs_i) = k$.

(4) A finite set *BE* of *boundary equations*:
A boundary equations is a quadrupel of the form $(i, x, j, \bar{x})$ where i and j are boundaries, x and $\bar{x}$ are dual variable bases, i is in $Col(x)$ and j is in $Col(\bar{x})$. Symbols $E, E_1, \ldots$ denote boundary equations.

The role of parts (1)-(3) of the definition will become clear in a moment when we have given the definition of a unifier of a generalized equation and when we show how word equations are translated into sets of generalized equations. In order to understand the role of the boundary equations (4), more background is needed. For the moment, imagine that a boundary equation $(i, x, j, \bar{x})$ expresses that the position of i in x corresponds to the position of j in $\bar{x}$.

For readers which are familiar with Jaffar's [Ja90] corresponding definition let us point out two differences: by (2), boundaries are naturally ordered and our generalized equations are completed in the sense of [Ja90]. The condition $Col(x) \neq Col(\bar{x})$ for non-empty variable bases x may be regarded as a normalization condition in the sense of [Ma77]. It will guarantee later that the "carrier"of a non-trivial generalized equation and its dual never have the same position.

Definition 1.2. Every assignment S of non-empty words $S(i, i+1) \in (\mathcal{C} \cup \mathcal{V})^+$ to the indecomposable columns of *GE* has — via concatenation — a unique extension which assigns a (non-) empty word to every (non-) empty column of *GE*. We identify S with this extension. S is a *unifier* of *GE* if three conditions are satisfied:

(i) $S(Col(bs_i)) = a$ for every constant base bs_i of type a $(a \in \mathcal{C})$,
(ii) $S(Col(x)) = S(Col(\bar{x}))$ for all variable bases x of *GE*,
(iii) $S(Left(x), i) = S(Left(\bar{x}), j)$ for every boundary equation $(i, x, j, \bar{x})$ in *GE*.

The *index* of S is the number $|S(1, M)|$, where M is the maximal boundary and $|W|$ denotes the length of the word W. The *exponent of periodicity* of S is the maximal number e such that $S(Col(x))$ may be represented in the form $S(Col(x)) = UV^eW$, V non-empty, for a variable base x of *GE*. S is a *solution* of *GE* if $S(1, M) \in \mathcal{C}^+$.

[5] i.e., a variable base with non-empty column.

Translation

The following example shows how word equations *WE* are translated into sets *ge*(*WE*) of generalized equations. We visualize one generalized equation which is assigned to the word equation WE_0 of the form $axxbyx == xayyxy$ with variables x and y:

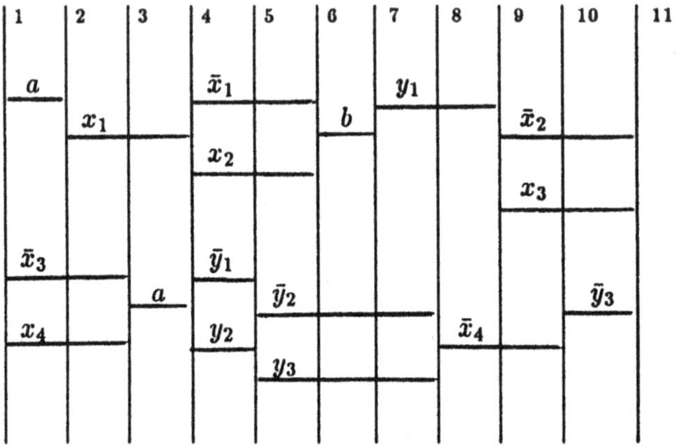

The vertical lines are the boundaries $1, 2, \ldots, 11$ of GE_0 which fix the relative extension of variables. GE_0 contains a certain "variant" of the left side $axxbyx$ of WE_0 in the upper part and a similar variant of the right side $xayyxy$ in the lower part.[6] In GE_0, multiple occurrences of the same symbol are formally distinguished. For this reason bases are introduced (horizontal lines). For the sake of simplicity we did not use distinct names for the two coefficient bases of type a in our figure. More important is the variable part. In word equations, a variable may have an arbitrary number of occurrences. In a generalized equation, every variable base has exactly two "dual" occurrences which are notationally distinguished by means of a bar "‾". By 1.2 (ii), dual bases have to get the same value. With this dualism it will be possible to transform a generalized equation without any enlargement of the number of bases. Exactly for this reason the variable dualism is introduced.[7]

When we translate WE_0 we must store the information that all columns of GE_0 which correspond to the four occurrences of x in WE_0 have to get the same value. We may only use pairs of dual variable bases. But we may also identify distinct variable bases by writing them into the same column. In our example every solution of GE_0 will assign the same value to $\bar{x}_1$ and to x_2, similarly to $\bar{x}_2$ and x_3 and also to $\bar{x}_3$ and x_4 because they have the same column. In combination with the equality of dual bases this will ensure that the four "x"-columns $(2, 4)$,

[6] The vertical position is, however, irrelevant — it was just chosen for the sake of readability.

[7] Later we shall see that this does not mean that we have avoided the "explosion of the data size" - generalized equation may start growing in another part.

(4,6), (9,11) and (1,3) will get the same value under an arbitrary solution. The same holds for the "y"-columns (7,9), (4,5), (5,8) and (10,11).

The remaining elements of $ge(WE_0)$, the set of all generalized equations corresponding to WE_0, only differ from the one given above in the relative position of bases. In order to preserve solvability the elements of $ge(WE_0)$ must represent every possible distribution for the relative length of the bases. The formal definition of the translation algorithm is not difficult and therefore omitted. We refer to [Ja90].

The following lemma summarizes some properties of the translation which are trivial but become important later. If S is a solution of the word equation WE, the exponent of periodicity of S is the maximal number e such that $S(x)$ may be represented in the form $S(x) = UV^eW$, V non-empty, for a variable x of WE.

Lemma 1.3. *There exists an algorithm which computes for every word equation WE_0 a finite set $ge(WE_0)$ of generalized equations with the following properties:*

(a) *WE_0 has a unifier with exponent of periodicity e if and only if some $GE \in ge(WE_0)$ has a unifier with exponent of periodicity e.*
(b) *The elements of $ge(WE_0)$ do not contain boundary equations. Every boundary is the right or left boundary of a base.*
(c) *For $GE \in ge(WE_0)$, the number of bases of GE does not exceed $2l(WE_0)$, where $l(WE_0)$ is the notational length[8] of WE_0.*

As in [Ja90], a generalized equation GE is called *trivially true*, if all variable bases of GE are empty and if GE has a unifier. GE is *true*, if it is trivially true or if all constant bases of GE have the same type and GE has a unifier. (If all constant bases of GE have the same type, then unifiability reduces to a set of length restrictions which may be represented by an existential formula of first-order arithmetic without multiplication (Presburger arithmetic). The validity of such formulas is decidable, see [Coo72]). GE is *trivially false* if two constant bases of distinct type have the same column. GE is *false*, if it is trivially false or if the generalized equation GE^a which we get when we assign the same type $a \in C$ to all constant bases has no unifier (in this case some inherent length restrictions cannot be satisfied). GE is trivial if it is either trivially true or trivially false.

Since empty variable bases cannot be involved in boundary equations it is trivial to decide the unifiability of GE if all variable bases are empty.

Lemma 1.4. *It is decidable whether a generalized equation is trivial (true, false).*

[8] i.e., the number of symbol occurrences.

2 Transformation — Notions

Suppose that *GE* is a non-trivial generalized equation. Let $l^\star$ denote the leftmost boundary among all left boundaries of non-empty variable bases. The *carrier* of *GE* is the largest[9] base among all variable bases with left boundary $l^\star$ (if there are several candidates, any may be chosen). The symbol x_c will be used to denote the carrier. The basic idea of the transformation procedure is to carry a part of the structure of $Col(x_c)$ to $Col(\bar{x}_c)$ and to erase a left part of *GE* afterwards. In general there are various ways how the structures of the two columns $Col(x_c)$ and $Col(\bar{x}_c)$ can be superposed and transformation is non-deterministic. In order to maintain unifiability downwards and upwards, all relevant information on identical subwords has to be preserved at a transformation step. As it turns out, it is possible to transport simultaneously the complete structure of $Col(x_c)$ up to a certain *critical boundary*. From now on, $(l^\star, r^\star)$ and $(l^{\star\star}, r^{\star\star})$ always denote the columns of x_c and $\bar{x}_c$ respectively.

Definition 2.1. The critical boundary of *GE* is the leftmost boundary among all left boundaries of variables bases y such $r^\star$ is in $Col(y)$, if such a base exists, and $r^\star$ in the other case. The symbol *cr* denotes the critical boundary.

Remark 2.2. In any non-trivial generalized equation *GE*, $l^\star < cr \leq l^{\star\star}$.

Up to x_c and $\bar{x}_c$, the bases and boundaries of *GE* will now be partitioned in three classes of superfluous objects, transport objects and fixed objects. The superfluous objects will be erased at a transformation step. The transport entites are carried to $Col(\bar{x}_c)$, the fixed entities keep their position. Roughly, the classification may be seen in the following figure, but details will become important later.

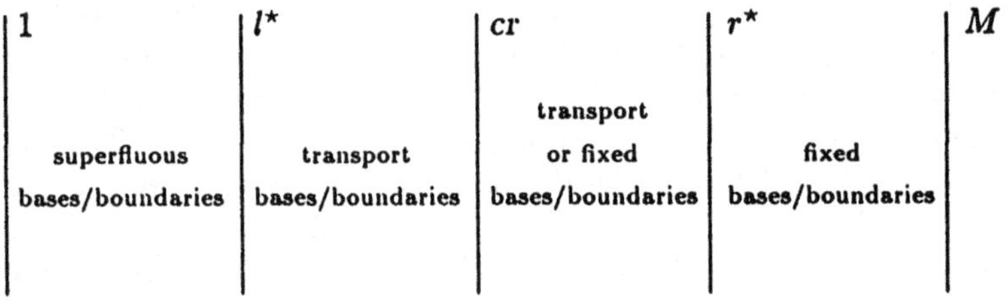

Definition 2.3.. The *transport bases* of *GE* are all bases $bs \neq x_c$ such that $l^\star \leq Left(bs) < cr$ and the empty bases with column (cr, cr). A base *bs* with $Col(bs) = (i, i)$, $i < l^\star$ is called *superfluous*. A base $bs \neq x_c, \bar{x}_c$ which is not superfluous and not a transport base is called a *fixed* base.

Note that *all* bases *bs* with $Left(bs) < l^\star$ are necessarily empty, by definition of the carrier.

[9] i.e., the one with the largest right boundary.

Remark 2.4. It will be frequently used that columns (i, j) of non-empty transport bases are always subcolumns of $(l^\star, r^\star)$ with $i < cr$. On the other hand, if (i, j) is the column of a fixed base, then $cr \leq i$ and $cr < j$.

Definition 2.5. A boundary i of *GE* is a *transport boundary* in three cases:

(i) if $l^\star < i \leq cr$,
(ii) if $cr < i < r^\star$ and $i = Right(x)$ where x is a transport base,
(iii) if $cr < i < r^\star$ and i occurs in a boundary equation $(i, x, j, \bar{x})$ or $(j, \bar{x}, i, x)$ where x is a transport base.

A boundary $i \leq l^\star$ is called *superfluous*. A boundary is *fixed* if it is neither superfluous nor a transport boundary.

When we carry the transport entities from $Col(x_c)$ to $Col(\bar{x}_c)$ we have to superpose the structures of both columns. The following definition excludes superpositions which are trivially wrong, contradicting informations about equal subparts of the two columns which are encoded in boundary equations of the form $(i, x_c, j, \bar{x}_c)$ or $(j, \bar{x}_c, i, x_c)$.

Definition 2.6. Let $l^{\star tr}, (l^\star+1)^{tr}, ..., r^{\star tr}$ be a sequence of symbols not occurring in *GE*, representing a copy of all boundaries between $l^\star$ and $r^\star$. An *extended print* is a linear order $\preceq$ on the set[10] $BD \cup \{l^{\star tr}, (l^\star+1)^{tr}, ..., r^{\star tr}\}$ satisfying the following conditions:

(i) $l^{\star tr} = l^{\star\star}, r^{\star tr} = r^{\star\star}$,
(ii) $\prec$ extends the natural order of BD and $k^{tr} \prec l^{tr}$ for $l^\star \leq k < l \leq r^\star$,
(iii) if $l^\star \leq i$ and $(i, j) = Col(bs_k)$ for a constant base, then i and j are consecutive with respect to $\preceq$. Similarly, if $l^\star \leq i < j \leq r^\star$ and $(i, j) = Col(bs_k)$ for a constant base, then also i^{tr} and j^{tr} are consecutive with respect to $\preceq$,
(iv) if $l^\star < i < r^\star$ and *GE* has a boundary equation $(i, x_c, j, \bar{x}_c)$ or $(j, \bar{x}_c, i, x_c)$, then $i^{tr} = j$,
(v) if x is a transport base, $\bar{x}$ a fixed base and if *GE* has a boundary equation $(i, x, j, \bar{x})$ or $(j, \bar{x}, i, x)$, then $i^{tr} = j$ iff $Left(x)^{tr} = Left(\bar{x})$ iff $Right(x)^{tr} = Right(\bar{x})$ (equalities with respect to $\preceq$).

A *print* is the restriction of an extended print to the set

$$BD' = \{cr, cr+1, ..., M\} \cup \{i_1^{tr}, \ldots, i_r^{tr}\}$$

where $\{i_1, ..., i_r\}$ is the set of all *transport* boundaries of *GE*.

Lemma 2.7. *The set of all prints of GE is finite and may effectively be computed.*

[10] For pure formalists: since elements of $M := BD \cup \{l^{\star tr}, (l^\star + 1)^{tr}, ..., r^{\star tr}\}$ may be identified with respect to $\preceq$, this linear order is formally an order on a partition of M.

Definition 2.8. The boundary equations of the form $E = (i, x, j, \bar{x}), x \neq x_c, \bar{x}_c$, are called *standard equations*. The *natural image* E' of E is the quadrupel which we get from E, replacing the entry i (entry j) by i^{tr} (by j^{tr}) if x (resp. $\bar{x}$) is a transport base and leaving it unchanged in the other case. Of course both bases x and $\bar{x}$ may be transport bases in which case both i and j have to be replaced to obtain the natural image. If $\preceq$ is a print, E' is *degenerate* with respect to $\preceq$ if the two boundaries occurring in E' coincide with respect to $\preceq$.

Remark 2.9. According to 2.4 and 2.6 (ii), $i^{tr} \prec cr^{tr}$ if i is the left boundary of a non-empty transport base, for any print $\preceq$.

3 Transformation — Algorithm

The following Transformation procedure assigns a finite set *Transf*(*GE*) (of non-false generalized equations) to every non-trivial generalized equation *GE*:

• **Transformation of** *GE* = (*BS*,*BD*,*Col*,*BE*).

Step 1: Compute the set of all prints for *GE*.

Step 2: For every print $\preceq$ of *GE* let $GE_{\preceq}$ be the generalized equation (BS', BD', Col', BE') with components as defined below. $GE_{\preceq}$ is — modulo a trivial renaming of boundaries — a generalized equation. *Transf1*(*GE*) is the set of all resulting structures $GE_{\preceq}$.

3.1 BS' is the set of all non-superfluous bases of GE.

3.2 BD' contains

3.2.1: all boundaries i, $cr \leq i$,

3.2.2: a new boundary i^{tr} for every transport boundary i of *GE*.

3.3 Col' is defined as follows:

3.3.1: $Col'(bs) = Col(bs)$ if *bs* is a fixed base of *GE*, with the exception described in 3.3.4,

3.3.2: $Col'(bs) = (Left(bs)^{tr}, Right(bs)^{tr})$ if *bs* is a transport base of *GE*, with the exception described in 3.3.4,

3.3.3: $Col'(x_c) = (cr, r^*)$ and $Col'(\bar{x}_c) = (cr^{tr}, r^{**})$,

3.3.4: If x is a non-empty variable base and if $Col'(x) = Col'(\bar{x})$ according to 3.3.1 and 3.3.2, then this value is corrected: $Col'(x) = Col'(\bar{x}) =$

$(Right'(x), Right'(x))$.

3.4: BE' contains:

3.4.1: every non-degenerate natural image E' of a standard equation E of GE,

3.4.2: all boundary equations of GE of the form $(i, x_c, j, \bar{x}_c)$ or $(j, \bar{x}_c, i, x_c)$ for $cr < i$,

3.4.3: a new equation $E' = (i, x_c, i^{tr}, \bar{x}_c)$ for every transport boundary $i > cr$.

Step 3: Erase all false elements of *Transf1*(GE). In the remaining structures, rename boundaries using natural numbers $1, \ldots, M'$, according to their order with respect to $\prec$. The resulting set is *Transf*(GE).

Example 3.5. To get a better picture of the algorithm it is useful to distinguish three levels of growing complication. In the first situation the boundary $r^\star$ is not inside the column of another base. Then $cr = r^\star$, the complete structure of $Col(x_c)$ is transported to $Col(\bar{x}_c)$ and $x_c, \bar{x}_c$ become empty bases. The transformation does not introduce any new boundary equation. In the following example, x is the carrier of GE, $r^\star = cr = 4$.

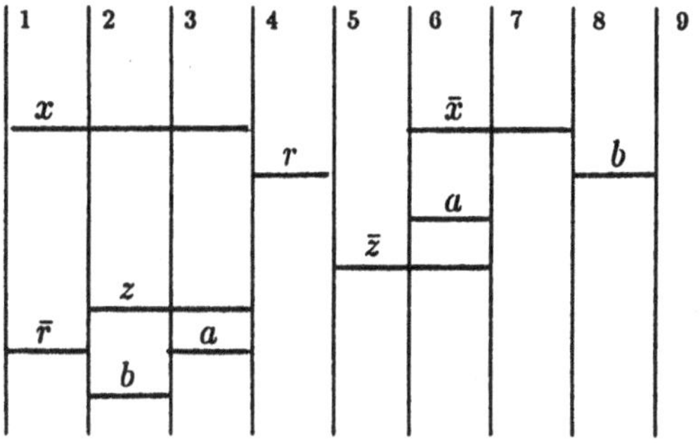

$(2, x, 7, \bar{x}), (3, z, 6, \bar{z})$

The first boundary equation is used to determine the new position 7 of 2^{tr} and erased afterwards. Here is one element of *Transf1*(GE). It is the only successor in this case — the empty variable bases x and $\bar{x}$ are omitted. Boundary names 1^{tr} and 4^{tr} are only added in order to facilitate the reading.

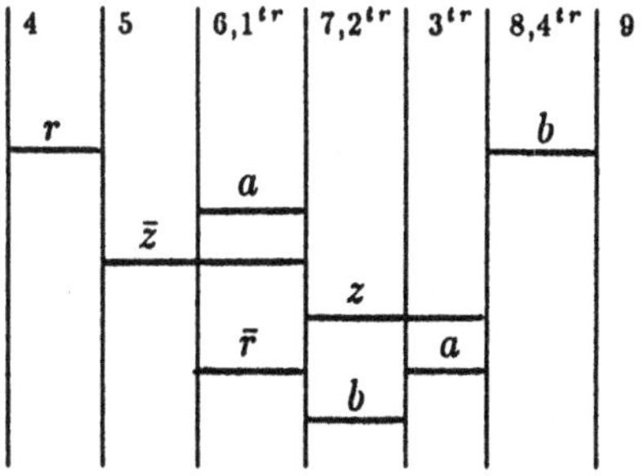

$(3^{tr}, z, 6, \bar{z})$

The corresponding element of *Transf*(GE) woud be obtained using standardized boundary names $1, \ldots, 7$.

The second typical situation occurs if there exists a base y, $\mathit{Left}(y) < r^\star$, which exceeds the carrier, but if there is no transport base whose right boundary falls into the column $(cr, r^\star)$. The subpart of $\mathit{Col}(x_c)$ up to the critical boundary is transported and cr becomes the new initial boundary. As a consequence of the second condition, no new boundary equations are introduced. The following generalized equation GE is an example — the carrier is x, $cr = 3$:

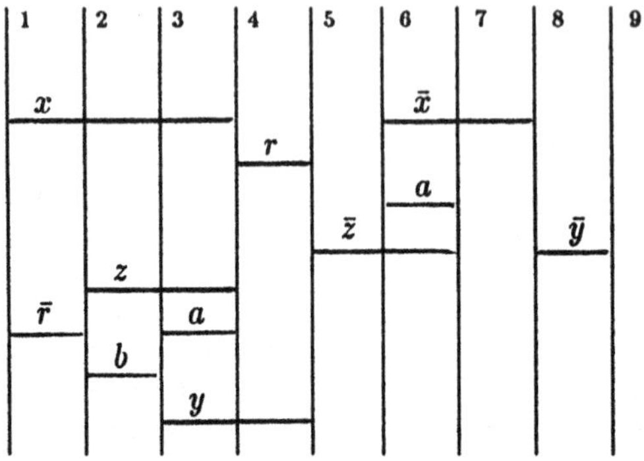

$$(2, x, 7, \bar{x}), (3, z, 6, \bar{z})$$

Again *Transf1*(GE) has only one element:

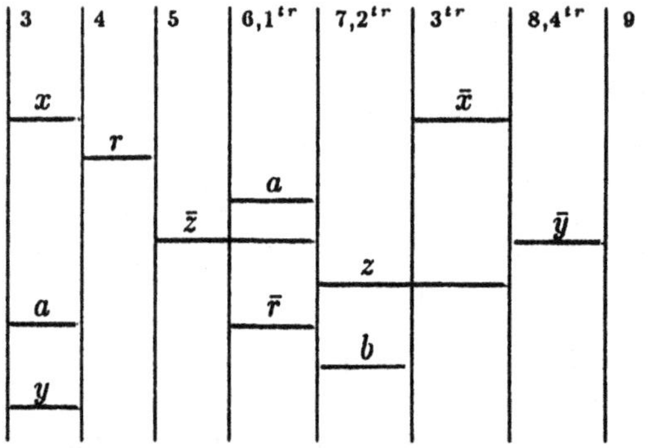

$$(3^{tr}, z, 6, \bar{z})$$

In the third and most complex situation we have a transport boundary between cr and $r^\star$. In our example it is the boundary $4 = \mathit{Right}(s)$.

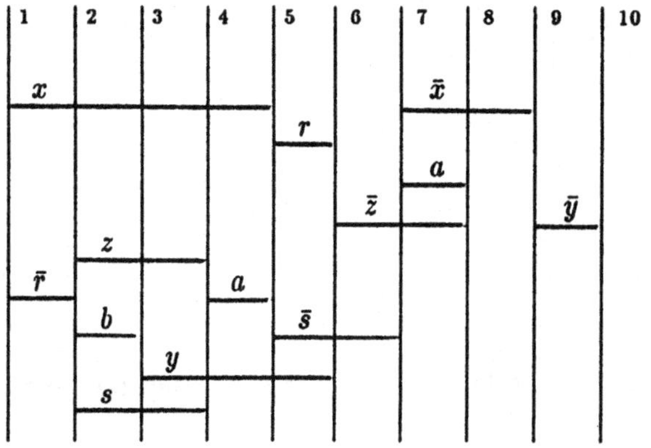

$$(2, x, 8, \bar{x}), (3, z, 7, \bar{z})$$

In this case we need a new boundary equation $(4, x, 4^{tr}, \bar{x})$ after the transformation in order to store all informations on identical subcolumns:

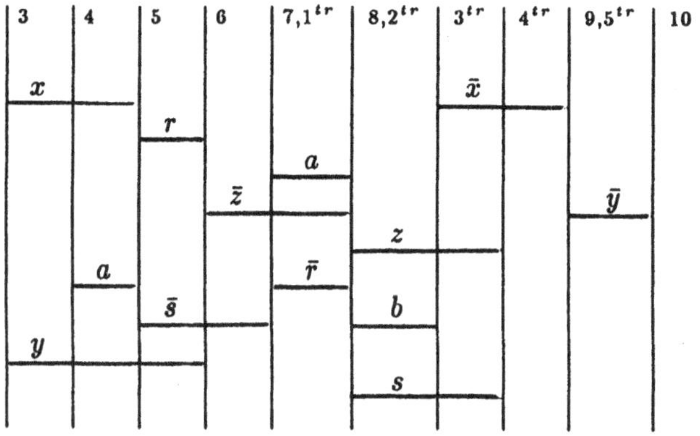

$(3^{tr}, z, 7, \bar{z}), (4, x, 4^{tr}, \bar{x})$

Remark 3.6. (a) Suppose that $GE' = GE_{\prec} \in$ *Transf*(GE). If the natural image of a boundary equation $E = (i, x, j, \bar{x})$ of GE is degenerate in GE', then we are in the case described in 3.3.4 and vice versa, by 2.6 (v).
(b) Note that the boundary equations of GE which are erased in GE' are exactly the degenerate boundary equations and the boundary equations of the form $(i, x_c, j, \bar{x}_c)$ or $(j, \bar{x}_c, i, x_c)$ where $i \leq cr$.

As a matter of fact, functions *Left'* and *Right'* may be defined in 3.3. In the following it is convenient to distinguish occurrences of the same base in GE and in $GE' := GE_{\prec}$ if their position is changed. We allow to write bs^{tr} for an occurrence of the transport base bs in GE', similarly we often write x'_c and $\bar{x}'_c$ for occurrences of x_c and $\bar{x}_c$ in GE'. We also allow to write l^{*tr} (r^{*tr}) for l^{**} (r^{**}).

Theorem 3.7. *For every print $\preceq$, the structure $GE_{\preceq}$ is - modulo renaming of boundaries - a generalized equation. The number N' of bases of $GE_{\preceq}$ does not exceed the number N of bases of GE.*

Proof: The only nontrivial part is to show that all elements E' of BE' are in fact boundary equations of GE'. If E' is the (non-degenerate) natural image of $E = (i, x, j, \bar{x})$, suppose, for example, that x is a transport base while $\bar{x}$ is fixed. Thus $E' = (i^{tr}, x^{tr}, j, \bar{x})$. Clearly j is in $Col'(\bar{x}) = Col(\bar{x})$, by 2.6 (ii) since E is a boundary equation. Again by condition (ii) of 2.6, i^{tr} is in $Col'(x^{tr})$ and E' is in fact a boundary equation. If E' has one of the types of 3.4.2 or 3.4.3, then it follows from $Col'(x'_c) = (cr, r^*)$ and $Col'(\bar{x}'_c) = (cr^{tr}, r^{**})$ that E' is a boundary equation, using 2.6 (ii) and, for 3.4.3, 2.6 (iv) to get $cr^{tr} \prec j$. ■

Before we continue the formal analysis of the transformation procedure we have to add a general remark: it is clear that the non-false elements of *Transf1*(GE) and the elements of *Transf*(GE) are identical modulo a trivial standardization of boundary names. We want to establish various results concerning

Transf(*GE*). For the proofs it is much more convenient to use the corresponding structures of *Transf1*(*GE*). Thus we shall henceforth ignore this notational distinction.

Theorem 3.8. (a) *If GE has a unifier S with index I and exponent of periodicity e, then Transf(GE) has an element GE′ which has a unifier S′ with index $I' < I$ and exponent of periodicity $e' \leq e$.*
(b) *If an element of Transf(GE) has a unifier, then GE is unifiable.*

Proof: (a) Let S be a unifier of *GE*. For all i in (l^*, r^*) and all j in (l^{**}, r^{**}) define

- $i^{tr} \prec j$ iff $S(l^*, i)$ is a proper prefix of $S(l^{**}, j)$,
- $i^{tr} = j$ iff $S(l^*, i) = S(l^{**}, j)$,
- $j \prec i^{tr}$ iff $S(l^{**}, j)$ is a proper prefix of $S(l^*, i)$.

Obviously $\preceq$ determines a unique extended print for *GE*. We may use the same symbol $\preceq$ for the corresponding print. We show that $GE' = GE_{\preceq}$ has a unifier S'. We have to consider the indecomposable columns of GE'. All such columns of the form $(i, i+1)$ which have an empty intersection with (l^{**}, r^{**}) are columns of *GE*. We define $S'(i, i+1) = S(i, i+1)$. There are at most the following four types of indecomposable subcolumns of (l^{**}, r^{**}) in GE':

(i) $(j, j+1)$ where $l^{**} \leq j < r^{**}$,

(ii) (i^{tr}, j) where i in (l^*, r^*) is a transport boundary and $l^{**} < j \leq r^{**}$,

(iii) (j, i^{tr}) where i in (l^*, r^*) is a transport boundary and $l^{**} \leq j < r^{**}$,

(iv) (i^{tr}, j^{tr}) where $l^* < i < j < r^*$ and both i and j are transport boundaries.

We define $S'(j, j+1) = S(j, j+1)$ in case (i), $S'(i^{tr}, j) = S(l^*, i)^{-1} S(l^{**}, j)$ in case (ii), $S'(j, i^{tr}) = S(l^{**}, j)^{-1} S(l^*, i)$ in case (iii) and $S'(i^{tr}, j^{tr}) = S(i, j)$ in case (iv). The following claim may be proved by induction on the number of indecomposable subcolumns. The technical proof is omitted:

Claim: For all common columns (i, j) of *GE* and GE': $S'(i, j) = S(i, j)$; for all columns of the form (i^{tr}, j^{tr}) of GE': $S'(i^{tr}, j^{tr}) = S(i, j)$.

With 3.3.1 and 3.3.2 it follows immediately that $S'(Col'(bs_i)) = S(Col(bs_i)) = a$ for all constant bases bs_i of GE' of type a, and that $S'(Col'(x)) = S'(Col'(\bar{x}))$ for all fixed and transport bases x of *GE*. Moreover, by the claim we get

$$S'(cr, r^*) = S(cr, r^*) = S'(cr^{tr}, r^{*tr}) = S'(cr^{tr}, r^{**})$$

and therefore $S'(Col'(x'_c)) = S'(Col'(\bar{x}'_c))$. Thus S' satisfies conditions (i) and (ii) of Definition 1.2.

Suppose now that E' is a boundary equation of GE'. Recall 3.4. We have to show that S' satisfies condition (iii) of 1.2 for E'. In the first case E' is the (non-degenerate) natural image of $E = (i, x, j, \bar{x}) \in BE$. Assume first that x is a transport base while $\bar{x}$ is fixed. Thus $E' = (i^{tr}, x^{tr}, j, \bar{x})$. By 3.3.2, the claim, 1.2 (iii) and 3.3.1 we have

$$\begin{aligned} S'(\mathit{Left}'(x^{tr}), i^{tr}) &= S'(\mathit{Left}(x)^{tr}, i^{tr}) = S(\mathit{Left}(x), i) \\ &= S(\mathit{Left}(\bar{x}), j) = S'(\mathit{Left}'(\bar{x}), j). \end{aligned}$$

Similarly condition (iii) of Definition 1.2 may be verified for the remaining subcases.

In the second case E' has the form $(i, x'_c, j, \bar{x}'_c)$ or $(j, \bar{x}'_c, i, x'_c)$ where $cr < i$ and where $E = (i, x_c, j, \bar{x}_c)$ or $(j, \bar{x}_c, i, x_c) \in BE$. Recall 3.3.3. Here $i^{tr} = j$, by 2.6 (iv). By the claim,

$$S'(cr, i) = S(cr, i) = S'(cr^{tr}, i^{tr}) = S'(cr^{tr}, j)$$

as demanded.

In the third case, $E' = (i, x'_c, i^{tr}, \bar{x}'_c)$ is new and $cr < i$, by 3.4.3. By the claim,

$$S'(cr^{tr}, i^{tr}) = S(cr, i) = S'(cr, i).$$

Thus condition (iii) of Definition 1.2 is always satisfied.

We have shown that S' is a unifier of GE'. Obviously the index of S' is strictly smaller than the index of S since $l^* < cr$. The exponent of periodicity e' of S' does not exceed the exponent of periodicity e of S since $S'(\mathit{Col}'(x))$ is always a suffix of $S(\mathit{Col}(x))$, for any variables base x of GE'.

(b) Assume now that S' is a unifier of the structure $GE' = GE_{\prec}$. All indecomposable columns $(i, i+1)$ of GE with $cr \leq i$ are columns of GE'. We define $S(i, i+1) = S'(i, i+1)$ in these cases. For the indecomposable columns $(i, i+1)$ of GE with $i + 1 \leq l^*$ we define $S(i, i+1) = a$ if $(i, i+1) = \mathit{Col}(bs_i)$ for a constant base of type a (remember that GE is nontrivial, hence a is unique). In the other cases, an arbitrary non-empty word $S(i, i+1)$ may be assigned to $(i, i+1)$ for $i < l^*$. The indecomposable columns $(l^*, l^*+1), \ldots, (cr-1, cr)$ of GE have images $(l^{**}, (l^*+1)^{tr}), \ldots, ((cr-1)^{tr}, cr^{tr})$ in GE' (3.2.2). We define $S(l^*, l^*+1) = S'(l^{**}, (l^*+1)^{tr}), \ldots, S(cr-1, cr) = S'((cr-1)^{tr}, cr^{tr})$. It follows immediately that $S'(i, j) = S(i, j)$ for all common columns (i, j) of GE' and GE. Furthermore, for all $l^* \leq i < j \leq cr$ we have $S(i, j) = S'(i^{tr}, j^{tr})$. Since $S'(cr, j) = S'(cr^{tr}, j^{tr})$ for all transport boundaries $j > cr$ of GE (compare 3.4.3, 3.3.3 and 1.2 (ii)) it is possible to prove that $S(i, j) = S'(i^{tr}, j^{tr})$ for arbitrary transport boundaries $i < j$ of GE. Similarly as in (a) it is now easy to verify that S is in fact a unifier of GE. ■

For word equation WE_0 the tree $T_{Mak}(WE_0)$ has WE_0 as top element and $ge(WE_0)$ as first level (compare lemma 1.3). For every $GE \in ge(WE_0)$, the downward tree is the unordered, finitely branching tree which results form iterated transformation.

Corollary 3.9. *WE_0 has a unifier if and only if $\mathcal{T}_{Mak}(WE_0)$ has a node which is labelled with a trivially true generalized equation.*

Proof: "*only if*": Suppose that WE_0 has a unifier. By 1.3 some element of $ge(GE)$ has a unifier, of index I, say. By 3.8 there is a downward branch in $\mathcal{T}_{Mak}(WE_0)$ labelled with unifiable generalized equations where the index decreases at every step. Since the index is non-negative, the length of this branch cannot exceed I. The generalized equation GE which labels the last node cannot have a non-empty variable base since otherwise the transformation algorithm would apply again. The rest is obvious.
"*if*": by Lemma 1.3 (a) and Theorem 3.8, with a trivial induction. ■

4 Proper Generalized Equations and Termination

In most cases $\mathcal{T}_{Mak}(WE_0)$ will be an infinite tree. The previous results show that it offers a correct and complete semi-decision procedure. In order to obtain termination we need two additional arguments. The first is simple: similarly as for the basic algorithm we may eliminate some branches by means of loop checking methods. Let us call two generalized equations GE and GE' *isomorphic* if the latter equation differs from the former only by means of a "consistent renaming of bases", i.e., if GE' may be obtained from GE by a bijection between the sets of bases which maps coefficient bases of the same type again into coefficient bases of the same type. Obviously we may stop a branch if we have found a generalized equation GE' which is isomorphic to a predecessor GE in the same branch.

Even with this pruning method we will get an infinite tree in general: generalized equations may have an arbitrary number of boundaries and boundary equations, thus there is an infinite number of non-isomorphic generalized equations even for a fixed number of bases. The second argument is much more difficult and may be regarded as the main idea behind Makanin's decidability result. It is based on the following theorem which was first proved in Bulitko [Bul]. A second version occurs in [Mak], recently Kościelski and Pacholski [KoP] found the bound which is used in the following formulation:

Theorem. *Let WE_0 be a word equation with notational length l. If WE_0 has any solution, then it has a solution where the exponent of periodicity e satisfies*

$$e \leq epp^{max}(l) = 2^{1.07l}.$$

Recall that translation into generalized equations and subsequent transformation steps preserve solvability under a given upper bound for the exponent of periodicity in the downward direction (1.3 and 3.8). Thus, for a mere decision procedure it suffices to consider in $\mathcal{T}_{Mak}(WE_0)$ the generalized equations which possible have a "tame" solution, i.e. a solution where the exponent of periodicity does not exceed $epp^{max}(l)$.

Makanin's main technical result — now adapted to the present terminology — was the proof that for the generalized equations *GE which are generated via transformation* the number of boundaries determines a lower bound for the exponent of periodicity of an arbitrary solution. If the number of boundaries is very large, *GE* cannot have a tame solution and may be treated as a failure leaf.

A similar result for arbitrary generalized equations can *not* be proved. The relevant properties which guarantee that a lower bound for the exponent of periodicity in terms of the number of boundaries may be given are captured by the concept of a *proper* generalized equation. This notion will be introduced below. In the longer version [Sc92] it is shown that all generalized equations in $\mathcal{T}_{Mak}(WE_0)$ are proper.

Let us continue with the decidability argument. Note that for all generalized equations in $\mathcal{T}_{Mak}(WE_0)$ the number N' of bases does not exceed the number $N = 2l(WE_0)$ where $l(WE_0)$ is the notational length of WE_0 (1.3 (c) and 3.7).

Theorem. *There exists a recursive function $NBD^{max}(N,b)$ such that every proper generalized equation with $N' \leq N$ bases and $M' \geq NBD^{max}(N,b)$ boundaries has only solutions S where the exponent of periodicity exceeds b.*

This is essentially Jaffar's Main Lemma ([Ja90], pg. 75). Now obviously the number of nonisomorphic proper generalized equations with N' bases and M' boundaries, where $N' \leq N$ and $M' \leq NBD^{max}(N, epp^{max}(l))$, is finite. Thus there is only a finite number of generalized equations to consider. Summarizing we arrive at the following

First Decision Procedure. Suppose the word equation WE_0 of length l is given. Let $epp^{max}(l)$ be the bound given in the theorem of Bulitko, Makanin and Kościelski-Pacholski. Translate WE_0 into $ge(WE_0)$, erasing false generalized equations. Iterate transformation. A node labelled with the generalized equation *GE* is a leaf in the following cases:

- *GE* is trivial: since *GE* is non-false, it is trivially true and solvable (success).
- *GE* is isomorphic to a predecessor equation (failure).
- If *GE* has $M' \geq NBD^{max}(N, epp^{max}(l))$ boundaries, where $N = 2l$ (failure).

Let $\mathcal{T}^{fin}_{Mak}(WE_0)$ be the resulting tree. It is finitely branching (2.7) and every path is of finite length. Thus $\mathcal{T}^{fin}_{Mak}(WE_0)$ is finite. WE_0 is unifiable if and only if $\mathcal{T}^{fin}_{Mak}(WE_0)$ has a leaf which is labelled with a trivially true generalized equation *GE*.

Proper Generalized Equations

If $E = (i, x, j, \bar{x})$ is a boundary equation of the generalized equation GE, then it is obviously possible to replace it by $E^{-1} = (j, \bar{x}, i, x)$ without affecting solvability. We shall say that *both* E and E^{-1} are *oriented versions* of E and *both* are oriented boundary equations of GE in this case. Throughout this subsection $GE = (BS, BD, Col, BE)$ denotes a nontrivial generalized equation. If not mentioned otherwise, all bases, boundaries and boundary equations are always from GE.

Definition 4.1. A *chain* of GE is a sequence $\pi = E_1, E_2, \ldots, E_m, w$ $(m > 0)$ where the E_l are oriented boundary equations of the form $(i_l, x_l, i_{l+1}, \bar{x}_l)$[11] $(1 \leq l \leq m)$ and w is a *witness*, i.e. a base $w = bs_j$ with $Right(w) = i_{m+1}$. The *leading base* of π is x_1.

Definition 4.2. Let $\pi = E_1, \ldots, E_m, w$ be a chain, suppose that E_l has the form $(i_l, x_l, i_{l+1}, \bar{x}_l)$ $(1 \leq l \leq m)$. To π we assign the word $\chi(\pi)$ of m symbols s_j in the alphabet $\{>, =, <\}$ defined by $Left(\bar{x}_j)\ s_j\ Left(x_{j+1})$ $(1 \leq j < m)$ and $Left(\bar{x}_m)\ s_m\ Left(w)$. The chain π is called *convex* if $\chi(\pi) \in \{>, =\}^* \circ \{<, =\}^*$. A convex chain π is *adaptive* if $\chi(\pi) \in \{<, =\}^*$.

For every unifier of GE, the values of the variable bases occurring in a convex chain π may be arranged to a "domino-tower" of the form indicated in the figure where parts which have vertical contact are identical. If π is adaptive, then the

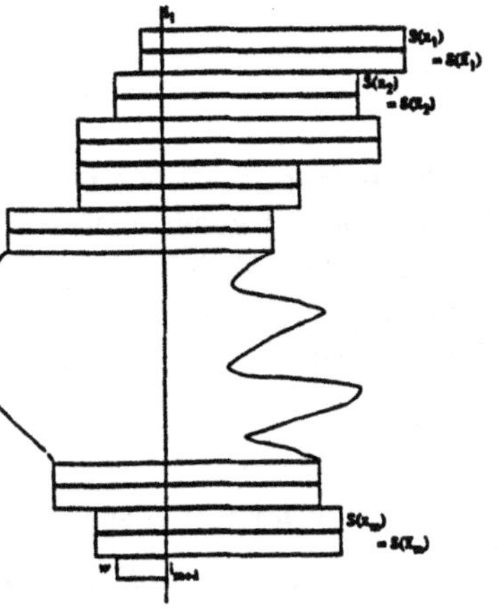

Fig. 1. A Domino-Tower built with solution components

upper part — the part which has a frontier growing leftwards — is empty. This

[11] Thus, the second boundary of E_l and the first boundary of E_{l+1} must always be identical.

means that we may put any additional "pair of stones" on top of the given tower without destroying the convex form.

Definition 4.3. The *left* (*right*) boundaries of GE are the boundaries i such that $Left(bs_j) = i$ ($Right(bs_j) = i$) for a base bs_j of GE. The *involved* boundaries are the boundaries i which occur in an oriented boundary equation $(i, x, j, \bar{x})$ of GE. A boundary i of GE is *abandoned* if it is neither left nor right nor involved. In the following $LB(GE)$ denotes the number of left boundaries of non-empty bases of GE and $AB(GE)$ the number of abandoned boundaries of GE.

Definition 4.4. The generalized equation GE with N bases and M boundaries is *proper*, if the following two conditions are satisfied:

(i) $LB(GE) + AB(GE) \leq N$.

(ii) For every boundary equation E of GE there exists a convex chain $E_1, E_2, ..., E_m, w$ where E_1 is an oriented version of E.

Let us briefly sketch the argument which shows that for proper generalized equations with $N' \leq N$ bases the number of boundaries determines a lower bound for the exponent of periodicity of an arbitrary solution. Condition (i) trivially implies that proper generalized equations with a large number of boundaries have a large number of boundary equations, for fixed number of bases. Condition (ii) implies that these boundary equations may be ordered to *long* convex chains π such that $\chi(\pi)$ contains a large number of symbols ">" or "<" (this is non-trivial). Such chains show that some solution component $S(Col(x_l))$ (here x_l is a variables base occurring in π) may be arranged to a high "domino-tower".

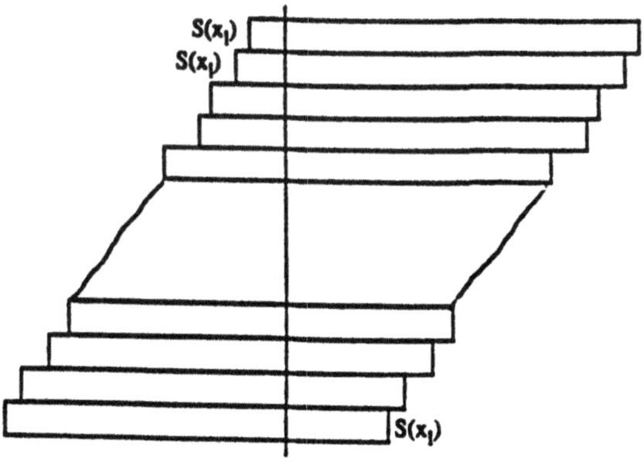

But such arrangements are only possible if $S(Col(x_l))$ has a large number of periodical, consecutive repetitions of the same nonempty subword. For details we refer to [Ma77, Ja90].

By 1.3 (b), all generalized equations in $ge(WE_0)$ are proper, for any word equation WE_0.

Theorem 4.5. *If GE is a non-trivial proper generalized equation, then all generalized equations in Transf(GE) are proper.*

See [Sc92] for a proof. It follows that all generalized equations occurring in $\mathcal{T}_{Mak}(WE_0)$ are proper.

5 Terminating Minimal and Complete Word Unification

The decision procedure of the preceding section could in principle be turned into an algorithm which computes a minimal and complete set of unifiers for a given word equation WE_0, terminating if this set is finite, using the same construction as in Jaffar [Jaf]. A detailed proof is however rather tedious. Following an idea of F. Baader we shall now describe an algorithm which does the same and is conceptually simpler. The basic idea is the following: we shall use the basic algorithm for word unification described in section 1 in order to generate a minimal and complete set of unifiers for a given word equation WE_0. With some additional amount of work these substitutions may be displayed at the successful leaves of $\mathcal{T}_{LP}(WE_0)$ (see below). The tree $\mathcal{T}_{LP}(WE_0)$ will be generated in breadth first manner, i.e., level for level, and we shall just collect all substitutions associated with successful leaves. As a matter of fact, each level contains only a finite number of word equations. Thus Makanin's decision procedure may be used as a subprocedure which decides for every level whether it contains a solvable equation or not. As soon as we have found a level where all equations are unsolvable we shall stop. We shall now show that the sequence of all substitutions which are displayed at the successful leaves of $\mathcal{T}_{LP}(WE_0)$ is in fact a minimal and complete set of unifiers for WE_0. Thus it is trivial that our algorithm generates such a set and terminates if this set is finite.

To be precise, let us recall several definitions. If S is a substitution and WE is a word equation of the form $W_1 == W_2$, then $S(WE)$ denotes the word equation $S(W_1) == S(W_2)$. Let $\mathcal{W} \subseteq \mathcal{V}$, let S, T be substitutions. Then S is more general than T with respect to $\mathcal{W}$, $S \leq T$ $(\mathcal{W})$, iff there exists a substitution R such that $R(S(x)) = T(x)$ for all $x \in \mathcal{W}$. A set Σ of unifiers for a word equation WE_0 is *complete* if for every unifier T of WE_0 there exists an $S \in \Sigma$ such that $S \leq T$ $(\mathcal{W}_0)$ where $\mathcal{W}_0$ is the set of variables occurring in WE_0. A complete set Σ of unifiers for WE_0 is *minimal* if it does not contain two elements $S_1 \neq S_2$ such that $S_1 \leq S_2$ $(\mathcal{W}_0)$. A stricter condition than minimality is ***disjointness***. A complete set Σ of unifiers for WE_0 is disjoint with respect to $\mathcal{W}_0$ if two distinct unifiers in Σ cannot be brought together: for all $S_1 \neq S_2 \in \Sigma$: there are no substitutions T_1, T_2 such that $T_1(S_1(x)) = T_2(S_2(x))$ for all $x \in \mathcal{W}_0$. $S \circ R$ denotes the product of two substitutions, S being applied first. $Var(WE)$ denotes the set of variables occurring in WE.

We shall now describe how unifiers may be displayed at successful leaves of $\mathcal{T}_{LP}(WE_0)$. For this purpose, let us associate with every word equation WE in $\mathcal{T}_{LP}(WE_0)$ the substitution S which is the product of all the substitutions S_i^{LP} which were applied at the transformation steps which led to WE. A convenient way to compute this substitution is to enrich WE_0 with a list $< x_1, ..., x_k >$ of its variables, representing the trivial substitution. At every transformation step the respective substitution S_i^{LP} is not only applied to the word equation, but also to the actual substitution list. Let Σ denote the set of all substitutions which are associated with successful leaves of $\mathcal{T}_{LP}(WE_0)$ in this way.

Theorem 6.1. *Σ is a disjoint and complete set of unifiers for WE_0.*

A proof was given in Siekmann's thesis [Si78], but the notation used there is more complicated. For the convenience of the reader we include a rather compact argument.
It is trivial that the elements of Σ are unifiers for WE_0. Let $\mathcal{W}_0$ denote the set of variables occurring in WE_0. The following lemma immediately implies that Σ is a complete set of unifiers for WE_0.

Lemma 6.2. *Every unifier T of WE_0 recursively defines a path π through $\mathcal{T}_{LP}(WE_0)$ with the following property: if WE is a word equation occurring in π with associated substitution S, then $S \leq T$ $(\mathcal{W}_0)$.*

Proof: The topmost node of π contains WE_0, and it is clear that Id (identity) satisfies $Id \leq T$ $(\mathcal{W}_0)$. Suppose now for the induction hypothesis that WE is in π with associated substitution S satisfying the condition of the lemma. Thus there exists a substitution R such that $R(S(z)) = T(z)$ for all $z \in \mathcal{W}_0$. For the induction step let us consider the case where WE has the head (a, x). To find the successor of WE with respect to T

(1) we apply S_1^{LP} if $R(x) = a$,

(2) we apply S_2^{LP} if $R(x) = aV$, with $V \in (\mathcal{V} \cup \mathcal{C})^+$.

This subcase analysis is complete: R unifies WE since $S \circ R$ is a unifier for WE_0 and WE is a suffix of $S(WE_0)$ which may be reached by iterated deletion of identical head-symbols form both sides. In the first case we define $T_1' : y \mapsto R(y)$ for $x \neq y \in Var(WE)$, in the second case we define $T_2' : x \mapsto V$ and $y \mapsto R(y)$ for $x \neq y \in Var(WE)$. Now T_1' and T_2' show that $S_i^{LP} \leq R$ (i = 1,2).
The substitution associated with the successor is $S \circ S_i^{LP}$ in case (i), i = 1,2. Let $z \in \mathcal{W}_0$. We have $T_i'(S_i^{LP}(S(z))) = R(S(z)) = T(z)$. Thus $S \circ S_i^{LP} \leq T$ $(\mathcal{W}_0)$. The proof for the situation where WE has head of type (x, y) is completely analogous. ■

If WE is any word equation in $\mathcal{T}_{LP}(WE_0)$ it is straightforward to show that the substitutions S_i^{LP} which may be applied are disjoint with respect to the variables occurring in the actual equation WE. However, in order to prove that Σ is a disjoint set of unifiers for WE_0 we have to show that the substitutions associated with word equations in distinct paths are disjoint with respect to the variables $\mathcal{W}_0$ of WE_0. For this purpose we shall introduce the notion of a D-preserving (disjointness-preserving) substitution:

Definition 6.3. A substitution S is *D-preserving* with respect to a set $\mathcal{W}$ of variables iff the following holds: for any two substitution T_1 and T_2 which are disjoint with respect to $\bigcup\{Var(S(x)); x \in \mathcal{W}\}$ the substitutions $S \circ T_1$ and $S \circ T_2$ are disjoint with respect to $\mathcal{W}$.

Lemma 6.4. *The transformation substitutions $S_i^{(LP)}$ are D-preserving with respect to the variables occurring in the actual word equation WE to be transformed $(i = 1, ..., 5)$.*

Proof: Let us treat the situation where WE has head of type (x, y). Let us consider S_4^{LP}. Let T_1 and T_2 be two substitutions. Assume that $S_4^{LP} \circ T_1$ and $S_4^{LP} \circ T_2$ are *not* disjoint with respect to $Var(WE)$. Then there exist substitutions R_1 and R_2 such that $R_1(T_1(S_4^{LP}(z))) = R_2(T_2(S_4^{LP}(z)))$ for all $z \in Var(WE)$. In particular,

$$R_1(T_1(S_4^{LP}(x))) = R_2(T_2(S_4^{LP}(x)))$$
$$R_1(T_1(S_4^{LP}(y))) = R_2(T_2(S_4^{LP}(y))).$$

Thus

$$R_1(T_1(yx))) = R_2(T_2(yx))$$
$$R_1(T_1(y))) = R_2(T_2(y))$$

and thus $R_1(T_1(x))) = R_2(T_2(x))$. Since for $z \neq x$ always $S_4^{LP}(z) = z$, this shows that T_1 and T_2 are not disjoint with respect to $Var(S_4^{LP}(WE)$. Thus S_4^{LP} is in fact D-preserving. The proof for S_5^{LP} is symmetric, the proof for S_3^{LP} is trivial, the proof for head (a, x) is analogous. ■

Lemma 6.5. *The set Σ is disjoint with respect to $\mathcal{W}_0$.*

Proof: The transformation substitutions S_i^{LP} which may be applied are disjoint with respect to the actual variable set. Suppose that S_1 and S_2 are substitutions associated with distinct successful leaves of $\mathcal{T}_{LP}(WE_0)$. A simple induction on the length of the common part of the paths leading to the respective leaves based on the preceding lemma shows that S_1 and S_2 are disjoint with respect to $\mathcal{W}_0$. ■

Improved Decision Procedure

Let us conclude with an improved decision procedure where some ideas from the basic algorithm are used for a pre-analysis of word equations which always simplifies the decision procedure described in section 4 and makes it even dispensable in some cases. We have to choose a slightly modified representation of a word equation.

Example 6.6. The word equation $axbzx == zczyyy$ is translated into the following matrix

$$\left|\begin{matrix} ax_1bz_1x_2 \\ z_2cz_3y_1y_2y_3 \end{matrix}\right| \left|\begin{matrix} x_1 \\ x_2 \end{matrix}\right| \left|\begin{matrix} y_1 \\ y_2 \\ y_3 \end{matrix}\right| \left|\begin{matrix} z_1 \\ z_2 \\ z_3 \end{matrix}\right|$$

representing the four multi-equations $ax_1bz_1x_2 == z_2cz_3y_1y_2y_3$ (principal multi-equation), $x_1 == x_2$, $y_1 == y_2 == y_3$ and $z_1 == z_2 == z_3$.

Note that every new indexed variable has exactly two occurrences. Thus the new structure lies half on the way between word equations and generalized equations. We may now apply the Lentin/Plotkin transformation strategy in order to resolve all columns with two lines only — it is simple to see that the number of symbols cannot grow! The first two successors are the following systems (simplifying the first system in the straightforward way):

$$\left|\begin{matrix} x_1bax_2 \\ cay_1y_2y_3 \end{matrix}\right| \left|\begin{matrix} x_1 \\ x_2 \end{matrix}\right| \left|\begin{matrix} y_1 \\ y_2 \\ y_3 \end{matrix}\right| \qquad \left|\begin{matrix} x_1bz_1x_2 \\ z_2cz_3y_1y_2y_3 \end{matrix}\right| \left|\begin{matrix} x_1 \\ x_2 \end{matrix}\right| \left|\begin{matrix} y_1 \\ y_2 \\ y_3 \end{matrix}\right| \left|\begin{matrix} z_1 \\ az_2 \\ z_3 \end{matrix}\right| .$$

Similar transformation steps are applied as long as there is any column with two lines left. We stop if a system is reached which is isomorphic to a predecessor in the same path. Eventually, when we reach a system where all columns have at least three lines, the matrix is translated into an equivalent set of generalized equations, introducing boundaries between all symbols occurring in a line of a column and choosing a linear order between the boundaries of the same column.

The use of such multi-equation systems has various advantages (see [Sc90] for a detailled discussion). For all word equations where every variable occurs at most twice the translation into generalized equations is completely avoided. Perhaps the most important point is the following: when the principal multi-equation is resolved, the number of additional transformation steps which lead to a unifier S cannot exceed the number $|X_1X_2 \ldots X_n|$ where $X_i = S(x_i)$ and $x_1, \ldots, x_n$ are the variables occurring in WE_0, due to the vertical orientation of non-principal columns. Thus the maximal number of such transformation steps is *independent from the number of occurrences of each variable*, in contrast to the situation in the first decision procedure. The resolution of principal multi-equation corresponds to the computation of the possible linear orders between boundaries for the structures in $gc(WE_0)$ in the standard approach. This amount of work cannot be avoided in either case.

References

[Ab87] H. Abdulrab, "Résolution d'équations sur les mots: Etude et implémentation LISP de l'algorithme de MAKANIN," Thèse de doctorat - Laboratoire d'informatique, Rouen 1987.

[AbPe89] H. Abdulrab, J.-P. Pécuchet, "Solving Word Equations," *J. Symbolic Computation* 8 (1989), pp. 499-521.

[Ah90] A.V. Aho, "Algorithms for Finding Patterns in Strings," in *Handbook of Theoretical Computer Schience* (J. van Leeuwen, Ed.) Elsevier Science Publishers 1990, pp. 256-300.

[Ba90] F. Baader, "Unification Theory," *Proceedings of the First International Workshop on Word Equations and Related Topics IWWERT '90*, (K.U. Schulz, Ed.) Tübingen 1990, Springer LNCS 572.

[Bu70] V.K. Bulitko, "Equations and Inequalities in a Free Group and a Free Semigroup," *Tul. Gos. Ped. Inst. Učen. Zap. Mat. Kafedr Vyp.2, Geometr. i Algebra*, (1970), pp. 242-252 (Russian).

[Col88] A. Colmerauer, "Final Specification for Prolog III," ESPRIT Ref. number P1219(1106), 1988.

[Coo72] D.C. Cooper, "Theorem Proving in Arithmetic without Multiplication," Machine Intelligence 7 (1972), pp. 82-95.

[Hm71] J.I. Hmelevskiĭ, "Equations in Free Semigroups", *Trudy Mat. Inst. Steklov*, Vol. 107, 1971; English translation: *Proc. Steklov Inst. of Mathematics*, Vol 107, 1971.

[Ja90] J. Jaffar, "Minimal and Complete Word Unification", *Journal of the ACM* **37**, No. 1, 1990, pp. 47-85.

[KoPa91] A. Kościelski, L. Pacholski, "Complexity of Makanin's Algorithms", Research Report, University of Wroclaw (1991); preliminary version: Complexity of Unification in Free Groups and Free Semigroups, *Proceedings 31st annual IEEE Symposium on Foundations of Computer Science*, Los Alamitos 1990, pp.824-829.

[Le72] A. Lentin, "Equations in Free Monoids", in *Automata Languages and Programming*, (M.Nivat, Ed.) North Holland Publishers, Amsterdam 1972, pp. 67-85.

[LeSc67] A. Lentin, M.P. Schützenberger, "A Combinatorial Problem in the Theory of Free Monoids", in *Proceedings of the University of North Carolina*, (1967) pp. 67-85.

[LiSi75] M. Livesey, J. Siekmann, "Termination and Decidability Results for String Unification", Essex University, 1975.

[Ma77] G.S. Makanin, "The problem of solvability of equations in a free semigroup", *Math. USSR Sbornik* **32**, 2 (1977), pp. 129-198.

[Ma80] G.S. Makanin, "Recognition of the Rank of Equations in a free semigroup", *Math. USSR Izvestija* Vol. 14 (1980) No. 3, pp. 499-545.

[Ma81] G.S. Makanin, "Equations in a free semigroup", *Amer. Math. Soc. Transl.* (2) Vol. 117 (1981).

[Pé81] J.P. Pécuchet, "Equations avec constantes et algorithme de Makanin", Thèse de doctorat, Laboratoire d' informatique, Rouen 1981.

[Pl72] G. Plotkin, "Building-in Equational Theories", *Machine Intelligence* 7 (1972), pp. 73-90.

[Sc90] K.U. Schulz, "Makanin's Algorithm - Two Improvements and a Generalization", (Habilitationsschrift), CIS-Report 91-39, University of Munich, also in *Proceedings of the First International Workshop on Word Equations and Related Topics IWWERT '90*, Tübingen 1990, Springer LNCS 572.

[Sc92] K.U. Schulz, "Word Unification and Transformation of Generalized Equations", to appear in the *Journal of Automated Reasoning*.

[Si78] J. Siekmann,"Unification and Matching Problems", Ph.D. Thesis, Essex University, Memo CSA-4-78, 1978.

[Si89] J. Siekmann,"Unification Theory: A Survey", in C. Kirchner (Ed.), *Special Issue on Unification, Journal of Symbolic Computation* **7**, 1989.

Unification in the Combination of Disjoint Theories

Peter Auer

Institut für Computergraphik, Technical University Vienna, A–1040, Austria

Abstract

We consider unifaction modulo some equational theory E: Given are terms $s, t \in \mathcal{T}(E)$ built from the signature $\Sigma(E)$ of E and from variables x in $\mathcal{V}$. A substitution unifies s, t if $\sigma(s) \equiv_E \sigma(t)$, i.e. $\sigma(s), \sigma(t)$ are equivalent modulo theory E.

In particular we give a unification algorithm for theories $E = E_1 \cup \cdots \cup E_N$ which are combinations of theories with disjoint signatures, $\Sigma(E_i) \cap \Sigma(E_j) = \emptyset$ for $i \neq j$. Our method works if for each theory E_i there exists a *restricted* unification algorithm:

> Given a set of equations $P = \{s_1 \asymp t_1, \ldots, s_m \asymp t_m\}$, a linear ordering $<$ of the variables in P, a set L of locked variables, the algorithm returns solutions σ with the following properties:
>
> - $\sigma(s_j) \equiv_{E_i} \sigma(t_j)$
> - x does not occur in $\sigma(y)$ if $y < x$
> - $\sigma(x) = x$ if $x \in L$.

No other restrictions are needed for the theories E_i.

1 Introduction

In the past unification algorithms for various equational theories have been constructed, for a comprehensive reference see [6]. Then there arose interest to combine these unification algorithms. This is particularly important if automated deduction systems or automated theorem provers are used to solve problems including function symbols which satisfy certain equational axioms. Attempts to combine unification algorithms were successful for equational theories satisfying certain conditions, [1], [2], [3], [4], [5], [7], [8]. In some sense our combination algorithm is similar to that presented in [5] and it is nearly equivalent to that of [2], which was developed earlier than ours but independently.

We start with some notations: Let $\mathcal{V}$ be the set of variables. An equational theory $E = \{s_i \equiv t_i : 1 \leq i \leq n\}$ consists of a set of axioms $s_i \equiv t_i$ where $s_i, t_i \in \mathcal{T}(E)$ are terms built from the signature $\Sigma(E)$ of E and variables in $\mathcal{V}$ (clearly $\mathcal{V} \cap \Sigma(E) = \emptyset$). A substitution is a mapping $\sigma : \mathcal{T}(E) \to \mathcal{T}(E)$ such that $\sigma(f(t_1, \ldots, t_n)) = f(\sigma(t_1), \ldots, \sigma(t_n))$

for all $f(t_1,\ldots,t_n) \in \mathcal{T}(E)$, $f \in \Sigma(E)$. Clearly σ is defined by the substitutions $\sigma(x)$ of the variables $x \in \mathcal{V}$. An equation is a pair from $\mathcal{T}(E) \times \mathcal{T}(E)$ denoted as $s \asymp t$. A substitution σ solves $s \asymp t$ (or unifies s,t) modulo E if $\sigma(s) \equiv_E \sigma(t)$ where $\equiv_E$ denotes equivalence modulo E. A substitution solves a set of equations P if it solves all equations in P. The set of variables appearing in a set of equations P is denoted by var(P). Clearly only the substitutions $\sigma(x)$ with $x \in$ var(P) are relevant for a solution σ of P. If a term s appears as subterm in term t then we write $s \ll t$ (note that $s = t \Rightarrow s \ll t$).

A unification problem $\mathcal{P}$ consists of a finite set of equations P and some additional restrictions which must be satisfied by the solutions of P. A set of solutions S of $\mathcal{P}$ is complete if for all solutions σ of $\mathcal{P}$ there is some $\tau \in S$ and some λ such that $\sigma(x) \equiv_E \lambda\tau(x)$ for all $x \in$ var(P). A unification algorithm A takes a unification problem $\mathcal{P}$ as input and satisfies one of the following conditions:

OUT1 A returns

- YES if there is a solution of $\mathcal{P}$
- NO if there is no solution of $\mathcal{P}$.

OUT2 A returns

- a solution σ of $\mathcal{P}$ if there is one
- NO otherwise.

OUT3 A returns a possibly infinite sequence $\sigma_1, \sigma_2, \ldots$ of solutions of $\mathcal{P}$ such that $\{\sigma_i : i \geq 1\}$ is a complete set of solutions of $\mathcal{P}$. If the sequence is finite A reports the end of the sequence.

OUT4 A returns a finite and complete set of solutions of $\mathcal{P}$.

Clearly an algorithm satisfying OUTi can be modified to satisfy OUTj for $1 \leq j \leq i \leq 4$. We say A is a unification algorithm for theory E if A solves finite sets of equations without any further restrictions.

Definition 1 (Consistent theory) *An equational theory is called consistent if $x \not\equiv_E y$ for $x, y \in \mathcal{V}$, $x \neq y$.*

If theory E is not consistent then $s \equiv_E t$ for all $s,t \in \mathcal{T}(E)$ and all substitutions solve any set of equations. Hence we will restrict ourselves to consistent theories.

Let $E = E_1 \cup \cdots \cup E_N$ be the union of equational theories with disjoint signatures, $\Sigma(E_i) \cap \Sigma(E_j) = \emptyset$ for $i \neq j$, $\Sigma(E) = \Sigma(E_1) \cup \cdots \cup \Sigma(E_N)$. We want to construct a unification algorithm for theory E from unification algorithms for theories $E_1, \ldots, E_N$. In the following sections we will derive such a combination algorithm. The results are presented in the final section.

2 Splitting the set of equations P

Since we want to solve a set of equations $P \subseteq \mathcal{T}(E) \times \mathcal{T}(E)$ using unification algorithms for $E_1, \ldots, E_N$, we have to split P into parts $P_1, \ldots, P_N$, $P_i \subseteq \mathcal{T}(E_i) \times \mathcal{T}(E_i)$. This can be done by introducing new variables in the following way: Assume there is a subterm $s = f(s_1, \ldots, s_n)$ in P with $s_k = g(t_1, \ldots, t_m)$ for some $1 \leq k \leq n$, $f \in \Sigma(E_i)$, $g \notin \Sigma(E_i)$, $i \in \{1, \ldots, N\}$. Then s is replaced by $f(s_1, \ldots, s_{k-1}, x, s_{k+1}, \ldots, s_n)$ where x is a new variable not appearing in P and the equation $x \asymp s_k$ is added to P obtaining P'. Clearly a solution $\hat{\sigma}$ of P' solves P, too. If σ is solution of P then

$$\hat{\sigma}(y) = \begin{cases} \sigma(y) & y \neq x \\ \sigma(s_k) & y = x \end{cases}$$

solves P'.

Repeatedly applying the above transformation we get $\hat{P} = P_1 \cup \cdots \cup P_N$, $P_i \subseteq \mathcal{T}(E_i) \times \mathcal{T}(E_i)$, such that σ solves P if and only if there is some $\hat{\sigma}$ which solves $\hat{P}$ with $\hat{\sigma}(x) = \sigma(x)$ for all $x \in \mathrm{var}(P)$. Hence we have

Lemma 1 *For each finite set of equations $P \subseteq \mathcal{T}(E) \times \mathcal{T}(E)$ we can construct a finite set $\hat{P}$ with $\hat{P} = P_1 \cup \cdots \cup P_N$, $P_i \subseteq \mathcal{T}(E_i) \times \mathcal{T}(E_i)$, such that*

$$\sigma \textit{ solves } P \iff \exists \hat{\sigma} : \hat{\sigma} \textit{ solves } \hat{P} \textit{ and } \hat{\sigma}(x) = \sigma(x)\ \forall x \in \mathrm{var}(P).$$

3 Constructing a solution of $\hat{P}$ from solutions of $P_1, \ldots, P_N$

For all $i \in \{1, \ldots, N\}$ let σ_i solve P_i modulo E_i. Let $\sigma_i(x) = x$ if $x \in \mathcal{V} - \mathrm{var}(\hat{P})$. From that solutions we want to construct a solution $\hat{\sigma}$ of $\hat{P} = P_1 \cup \cdots \cup P_N$ modulo E. Clearly $\lambda_i \sigma_i$ solves P_i modulo E for all substitutions λ_i. Hence we have a solution $\hat{\sigma}$ of $\hat{P}$ if we find substitutions $\lambda_1, \ldots, \lambda_N$ such that $\lambda_1 \sigma_1 = \cdots = \lambda_N \sigma_N = \hat{\sigma}$. If σ_i is idempotent then $\hat{\sigma}\sigma_i = \lambda_i \sigma_i \sigma_i = \lambda_i \sigma_i = \hat{\sigma}$. Therefore we are looking for some $\hat{\sigma}$ with $\hat{\sigma}\sigma_i = \hat{\sigma}$ for all $i = 1, \ldots, N$.

Such a $\hat{\sigma}$ does not exist in general: Consider for example $\sigma_1(x) = f(s)$, $\sigma_2(x) = g(s)$. Then $\hat{\sigma}\sigma_1(x) = f(\hat{\sigma}(s)) \neq g(\hat{\sigma}(s)) = \hat{\sigma}\sigma_2(x)$. Hence we assume that $\sigma_i(x) \neq x$ for at most one $i \in \{1, \ldots, N\}$ and define

$$\sigma_0(x) = \begin{cases} \sigma_i(x) & \sigma_i(x) \neq x \\ x & \sigma_1(x) = \cdots = \sigma_N(x) = x. \end{cases}$$

Since $\hat{\sigma}\sigma_0(x) = \hat{\sigma}\sigma_i(x)$ for some $i \in \{1, \ldots, N\}$ we have $\hat{\sigma}\sigma_0 = \hat{\sigma} \Leftrightarrow \hat{\sigma}\sigma_i = \hat{\sigma}\ \forall i \in \{1, \ldots, N\}$. Hence $\hat{\sigma} = \sigma_0^K$ is a possible choice for $\hat{\sigma}$ if $\sigma_0^{K+1} = \sigma_0^K$. Assume there is some $x_0 \in \mathrm{var}(\hat{P})$ with $\sigma_0^k(x_0) \neq \sigma_0^{k+1}(x_0)$ for all $k \geq 0$. Then there must be a $x_1 \ll \sigma_0(x_0)$, $\sigma_0(x_0) \neq x_0$, $x_1 \in \mathrm{var}(\hat{P})$ with $\sigma_0^k(x_1) \neq \sigma_0^{k+1}(x_1)$ for all $k \geq 0$. Hence we have an infinite sequence $x_0, x_1, x_2, \ldots$ of variables from $\mathrm{var}(\hat{P})$ with $x_{i+1} \ll \sigma_0(x_i)$, $\sigma_0(x_i) \neq x_i$. Since there are only finitely many distinct variables in the sequence, there must be a circle

$$y_1 \ll \sigma_0(y_0), y_2 \ll \sigma_0(y_1), \ldots, y_m \ll \sigma_0(y_{m-1}), y_0 \ll \sigma_0(y_m) \tag{1}$$

with $\sigma_0(y_i) \neq y_i$. If there is a partial ordering of the variables with

$$x \ll \sigma_0(y) \Rightarrow x < y \vee \sigma_0(y) = y$$

then (1) implies $y_0 > y_1 > y_2 > \cdots > y_m > y_0$ which is a contradiction. Such a partial ordering exists for σ_0 if it exists for σ_i, $i = 1, \ldots, N$ since $x \ll \sigma_0(y)$ implies $x \ll \sigma_i(y)$ for some $i \in \{1, \ldots, N\}$. Summing up we have

Lemma 2 *Let $P_1, \ldots, P_N$ be sets of equations with $P_i \subseteq \mathcal{T}(E_i) \times \mathcal{T}(E_i)$. If σ_i is solution of P_i modulo E_i such that*

- *$\sigma_i(x) = x$ for $x \notin \mathrm{var}(P_1 \cup \cdots \cup P_N)$*
- *$\sigma_i(x) \neq x$ for at most one $i \in \{1, \ldots, N\}$*
- *there is a partial ordering $<$ of $\mathrm{var}(P_1 \cup \cdots \cup P_N)$ with*

  $$x \ll \sigma_i(y) \Longrightarrow \sigma_i(y) = y \vee x < y$$

  for all $i \in \{1, \ldots, N\}$

then there exists a K such that $\sigma_0(\sigma_1, \ldots, \sigma_N)^K$ solves $P_1 \cup \cdots \cup P_N$ and $\sigma_0^K = \sigma_0^{K+1}$.

4 Constructing solutions $\sigma_1, \ldots, \sigma_N$ from a solution of $\hat{P}$

In this section we construct from a solution σ of $\hat{P}$ solutions $\sigma_1, \ldots, \sigma_N$ of $P_1, \ldots, P_N$ which satisfies the conditions of lemma 2. At first we have to find some $\hat{\sigma}$ which solves P_i modulo E_i for all $i \in \{1, \ldots, N\}$. We construct $\hat{\sigma}$ by normalizing σ. We define a linear ordering $\prec$ on $\mathcal{T}(E)$ with the following properties:

- There is a minimal term $v_0 \in \mathcal{V} - \mathrm{var}(\hat{P})$ in respect to $\prec$.
- $s \prec f(t_1, \ldots, t_{k-1}, s, t_{k+1}, \ldots, t_n)$.
- $s_1 \prec s_2$ implies $f(t_1, \ldots, t_{k-1}, s_1, t_{k+1}, \ldots, t_n) \prec f(t_1, \ldots, t_{k-1}, s_2, t_{k+1}, \ldots, t_n)$.

Furthermore we define for all $t \in \mathcal{T}(E)$

$$R_{E_i}(t) = min_{\prec}\{s : s \equiv_{E_i} t\}, \; i \in \{1, \ldots, N\},$$

$$R(t) = min_{\prec}\{s : s \equiv_E t\}.$$

With some effort (see [1] for a complete proof) we can show

Proposition 1
$R(x) = x$ for all $x \in \mathcal{V}$,
$R(f(t_1, \ldots, t_n)) = R_{E_i}(f(R(t_1), \ldots, R(t_n)))$ if $f \in \Sigma(E_i)$.

Then $\hat{\sigma}$ is defined as $\hat{\sigma}(x) = R(\sigma(x))$ for all $x \in \mathcal{V}$. To prove that $\hat{\sigma}$ solves P_i modulo E_i we show that $R(\sigma(t)) = R_{E_i}(\hat{\sigma}(t))$ for all $t \in \mathcal{T}(E_i)$. Then we have for all $s \asymp t \in P_i$ that $R_{E_i}(\hat{\sigma}(s)) = R(\sigma(s)) = R(\sigma(t)) = R_{E_i}(\hat{\sigma}(t))$ and hence $\hat{\sigma}(s) \equiv_{E_i} \hat{\sigma}(t)$.

Proposition 2 *If $t \in \mathcal{T}(E_i)$ then $R(\sigma(t)) = R_{E_i}(\hat{\sigma}(t))$.*

Proof. The proof is by induction on the term depth of t. If $t \in \mathcal{V}$ then $R_{E_i}(\hat{\sigma}(t)) = R_{E_i}(R(\sigma(t))) = R(\sigma(t))$. Otherwise

$$\begin{aligned} R(\sigma(f(t_1,\ldots,t_n))) &= R_{E_i}(f(R(\sigma(t_1)),\ldots,R(\sigma(t_n)))) \\ &= R_{E_i}(f(R_{E_i}(\hat{\sigma}(t_1)),\ldots,R_{E_i}(\hat{\sigma}(t_n)))) \\ &= R_{E_i}(f(\hat{\sigma}(t_1),\ldots,\hat{\sigma}(t_n))) \\ &= R_{E_i}(\hat{\sigma}(f(t_1,\ldots,t_n))). \end{aligned}$$

□

To get $\sigma_1,\ldots,\sigma_N$ we have to split $\hat{\sigma}$ in an appropriate way. Unfortunately this is not possible in general. Consider for example

$$\begin{aligned} P_1 &= \{f(x_1) \asymp f(x_2), f(x_1) \asymp f(x_3)\} \\ P_2 &= \{g(x_1) \asymp g(x_2), g(x_1) \asymp g(x_3)\} \end{aligned}$$

and $\hat{\sigma}(x_1) = \hat{\sigma}(x_2) = \hat{\sigma}(x_3)$. Clearly σ_i must satisfy $\sigma_i(x_1) = \sigma_i(x_2) = \sigma_i(x_3)$ and hence it is not possible that $\sigma_1(x_j) = x_j$ or $\sigma_2(x_j) = x_j$ for all $j = 1,2,3$.

To overcome this problem we use a variable identification

$$\nu(x) = \min_{\prec}\{y \in \text{var}(\hat{P}) : \hat{\sigma}(y) = \hat{\sigma}(x)\}$$

for $x \in \text{var}(\hat{P})$. Since $\hat{\sigma}\nu = \hat{\sigma}$ $\hat{\sigma}$ solves $\nu(P_i)$ modulo E_i and we construct $\sigma_1,\ldots,\sigma_N$ such that σ_i solves $\nu(P_i)$, too. We use the following sets of transformation rules

$$\rho_i = \{\hat{\sigma}(x) \to \nu(x) : x \in \text{var}(\hat{P}), \hat{\sigma}(x) \neq f(t_1,\ldots,t_n), f \in \Sigma(E_i)\}.$$

The application $t|\rho$ of a set of rules is defined as

- $\exists t \to s \in \rho \Rightarrow t|\rho = s$
- $\nexists t \to s \in \rho \wedge t \in \mathcal{V} \Rightarrow t|\rho = t$
- $\nexists t \to s \in \rho \wedge t = f(t_1,\ldots,t_n) \Rightarrow t|\rho = f(t_1|\rho,\ldots,t_n|\rho)$.

Since ν identifies variables x with the same substitution $\hat{\sigma}(x)$, the transformations ρ_i are well-defined and σ_i is given by $\sigma_i(x) = \hat{\sigma}(x)|\rho_i$ for all $x \in \mathcal{V}$. To prove that σ_i solves $\nu(P_i)$ modulo E_i we show that $\hat{\sigma}(s) \equiv_{E_i} \hat{\sigma}(t)$ implies $\hat{\sigma}(s)|\rho_i \equiv_{E_i} \hat{\sigma}(t)|\rho_i$ and that $\hat{\sigma}(s)|\rho_i = \sigma_i\nu(s)$ for all $s,t \in \mathcal{T}(E_i)$. Then $s \asymp t \in P_i$ implies $\sigma_i\nu(s) = \hat{\sigma}(s)|\rho_i \equiv_{E_i} \hat{\sigma}(t)|\rho_i = \sigma_i\nu(t)$.

Proposition 3 $\hat{\sigma}(s)|\rho_i = \sigma_i\nu(s)\quad \forall s \in \mathcal{T}(E_i)$.

Proof. If $s \in \mathcal{V}$ then $\hat{\sigma}(s)|\rho_i = \hat{\sigma}(\nu(s))|\rho_i = \sigma_i(\nu(s))$ by definition. By induction on the term depth of s and by the fact that ρ_i includes no rule $f(\cdots) \to x$ with $f \in \Sigma(E_i)$ we get $\hat{\sigma}(f(t_1,\ldots,t_n))|\rho_i = f(\hat{\sigma}(t_1),\ldots,\hat{\sigma}(t_n))|\rho_i = f(\hat{\sigma}(t_1)|\rho_i,\ldots,\hat{\sigma}(t_n)|\rho_i) = f(\sigma_i\nu(t_1),\ldots,\sigma_i\nu(t_n)) = \sigma_i\nu(f(t_1,\ldots,t_n))$.

□

Proposition 4 $\hat{\sigma}(s) \equiv_{E_i} \hat{\sigma}(t) \Rightarrow \hat{\sigma}(s)|\rho_i \equiv_{E_i} \hat{\sigma}(t)|\rho_i \quad \forall s,t \in \mathcal{T}(E_i)$.

Proof. The proof is by induction on the equivalence relation $\equiv_{E_i}$. Assume there is some $u \equiv v \in E_i$ and a substitution λ with $\hat{\sigma}(s) = \lambda(u)$, $\hat{\sigma}(t) = \lambda(v)$. Let $\lambda_i(x) = \lambda(x)|\rho_i$. Then by the proof of proposition 3 $\lambda_i(u) = \lambda(u)|\rho_i$ and hence $\hat{\sigma}(s)|\rho_i = \lambda(u)|\rho_i = \lambda_i(u) \equiv_{E_i} \lambda_i(v) = \lambda(v)|\rho_i = \hat{\sigma}(t)|\rho_i$. Now assume $\hat{\sigma}(s) = f(s_1,\ldots,s_n)$, $\hat{\sigma}(t) = f(t_1,\ldots,t_n)$, $s_j \equiv_{E_i} t_j$. If $s,t \in \mathcal{V}$ then by definition of $\hat{\sigma}$ $\hat{\sigma}(s) = \hat{\sigma}(t)$ and hence $\hat{\sigma}(s)|\rho_i = \hat{\sigma}(t)|\rho_i$. If $s,t \notin \mathcal{V}$ then $s = f(u_1,\ldots,u_n)$, $t = f(v_1,\ldots,v_n)$, $f \in \Sigma(E_i)$, and by induction $\hat{\sigma}(s)|\rho_i = f(\hat{\sigma}(u_1)|\rho_i,\ldots,\hat{\sigma}(u_n)|\rho_i) \equiv_{E_i} f(\hat{\sigma}(v_1)|\rho_i,\ldots,\hat{\sigma}(v_n)|\rho_i) = \hat{\sigma}(t)|\rho_i$. If $s \notin \mathcal{V}$ and $t \in \mathcal{V}$ then there is a variant of $\hat{\sigma}$ such that $\hat{\sigma}(t) = \hat{\sigma}(f(v_1,\ldots,v_n))$ where $v_i \in \mathcal{V}$.

□

With the above definitions we have

Lemma 3 *If σ is a solution of $P_1 \cup \cdots \cup P_N$ modulo E with $\sigma\sigma = \sigma$ and $\sigma(x) = x$ if $x \notin P_1 \cup \cdots \cup P_N$ then for all $i \in \{1,\ldots,N\}$ σ_i is a solution of $\nu(P_i)$ modulo E_i. Furthermore*

- *$\sigma_i\nu(x) \neq \nu(x)$ for at most one $i \in \{1,\ldots,N\}$*
- *there is a partial ordering $<$ of $\mathrm{var}(\nu(P_1 \cup \ldots \cup P_N))$ such that $x \ll \sigma_i(y)$ implies $\sigma_i(y) = y$ or $x < y$ for all $i \in \{1,\ldots,N\}$*
- *$\sigma(x) \equiv_E \hat{\sigma}(x)$ for all $x \in \mathcal{V}$ and $\hat{\sigma}\sigma_i = \hat{\sigma}$ for all $i \in \{1,\ldots,N\}$.*

Proof. By the above constructions σ_i solves P_i modulo E_i and $\hat{\sigma} \equiv_E \sigma$. By definition $\hat{\sigma}(\nu(x)) = \nu(x)$ for all $x \notin \mathrm{var}(\hat{P})$. If $x \in \mathrm{var}(\hat{P})$ and $\nu(x) \neq \sigma_i(\nu(x)) = \hat{\sigma}(\nu(x))|\rho_i$ then $\hat{\sigma}(\nu(x)) = f(t_1,\ldots,t_N)$ with $f \in \Sigma(E_i)$. Hence $\sigma_i\nu(x) \neq \nu(x)$ for at most one $i \in \{1,\ldots,N\}$.

To prove that $\hat{\sigma}\sigma_i = \hat{\sigma}$ we show that $\hat{\sigma}\hat{\sigma} = \hat{\sigma}$ and $\hat{\sigma}(s|\rho_i) = \hat{\sigma}(s)$. Then $\hat{\sigma}\sigma_i(x) = \hat{\sigma}(\hat{\sigma}(x)|\rho_i) = \hat{\sigma}\hat{\sigma}(x) = \hat{\sigma}(x)$. Clearly $\tau\tau = \tau$ if and only if for all $x,y \in \mathcal{V}$ $x \ll \tau(y)$ implies $\tau(x) = x$. If $x \ll \hat{\sigma}(y) = R(\sigma(y))$ then $x \ll \sigma(y)$ or $x = v_0$ (see the definition of $\prec$). Hence $\hat{\sigma}(x) = R(\sigma(x)) = R(x) = x$. The proof of $\hat{\sigma}(s|\rho_i) = \hat{\sigma}(s)$ is by induction on the definition of $s|\rho_i$. If $s \to x \in \rho_i$ then $\hat{\sigma}(x) = s$ and $\hat{\sigma}(s|\rho_i) = \hat{\sigma}(x) = \hat{\sigma}\hat{\sigma}(x) = \hat{\sigma}(s)$. Otherwise let $s \in \mathcal{V}$. Then $\hat{\sigma}(s|\rho_i) = \hat{\sigma}(s)$. If $s = f(s_1,\ldots,s_n)$ then $\hat{\sigma}(s|\rho_i) = \hat{\sigma}(f(s_1|\rho_i,\ldots,s_n|\rho_i)) = f(\hat{\sigma}(s_1),\ldots,\hat{\sigma}(s_n)) = \hat{\sigma}(s)$.

The partial ordering is defined as $x < y \Leftrightarrow \hat{\sigma}(x) \ll \hat{\sigma}(y)$, $\hat{\sigma}(x) \neq \hat{\sigma}(y)$. Then

$$\begin{aligned} x \ll \sigma_i(y) &\Rightarrow \hat{\sigma}(x) \ll \hat{\sigma}\sigma_i(y) = \hat{\sigma}(y) \\ &\Rightarrow x < y \vee \hat{\sigma}(x) = \hat{\sigma}(y) \\ &\Rightarrow x < y \vee x = y \end{aligned}$$

since $x,y \in \mathrm{var}(\nu(\hat{P}))$.

□

5 The combination algorithm

From sections 3 and 4 we know that we need restricted unification algorithms for the theories $E_1, \ldots, E_N$ where a restricted unification algorithm is defined as follows.

Definition 2 (Restricted unification algorithm) *Let E_i be some equational theory. Then a restricted unification algorithm for theory E_i takes as input a triple $\mathcal{P} = (P, <, L)$ where $P \subseteq \mathcal{T}(E_i) \times \mathcal{T}(E_i)$ is a finite set of equations, $<$ is a linear ordering of* var(P), *$L \subseteq$ var(P), and satisfies one of the conditions* OUT1–OUT4. *A substitution σ solves $\mathcal{P} = (P, <, L)$ if*

- *σ solves P modulo E_i*
- *$x \ll \sigma(y)$ implies $x < y$ or $\sigma(y) = y$*
- *$\sigma(x) = x$ for all $x \in (\mathcal{V} - \text{var}(P)) \cup L$.*

In the following combination algorithms A for theory E satisfying OUT1–OUT4 are presented, relying on restricted unification algorithms $A_1, \ldots, A_N$ for theories $E_1, \ldots, E_N$. Algorithm A takes a finite set of equations $P \subseteq \mathcal{T}(E) \times \mathcal{T}(E)$ as input and produces solutions of P. The correctness proof of the combination algorithm is refered to the next section.

5.0 Preprocessing

1. Construct $\hat{P}$, $\hat{P} = P_1 \cup \cdots \cup P_N$, $P_i \subseteq \mathcal{T}(E_i) \times \mathcal{T}(E_i)$, as in section 2.

2. Let

$$\begin{aligned} \mathcal{L} = \{ \quad & (\nu, <, L_1, \ldots, L_N) : \\ & \nu : \text{var}(\hat{P}) \to \text{var}(\hat{P}), \nu\nu = \nu, \\ & < \text{ is a linear ordering of } \text{var}(\nu(\hat{P})), \\ & L_i \subseteq \text{var}(\nu(P_i)), \\ & (\text{var}(\nu(\hat{P}_i)) - L_i) \cap (\text{var}(\nu(\hat{P}_i)) - L_j) = \emptyset \quad \forall i \neq j \}. \end{aligned}$$

3. If σ_i is solution of $(\nu(P_i), <, L_i)$, $i = 1, \ldots, N$, then let

$$\sigma_0(x) = \sigma_0(\sigma_1, \ldots, \sigma_N)(x) = \begin{cases} \mu_i\sigma_i(x) & \sigma_i(x) \neq x \\ x & \sigma_1(x) = \cdots = \sigma_N(x) = x \end{cases}$$

   where $\mu_i : \mathcal{V} - L_i \to \mathcal{V} - \text{var}(\nu(\hat{P}))$ with $\mu_i(x) \neq \mu_i(y)$ for $x \neq y$ and $\text{var}(\mu_i\sigma_i(P_i)) \cap \text{var}(\mu_j\sigma_j(P_j)) \subseteq L_i \cap L_j$ for all $i, j \in \{1, \ldots, N\}$.

Remark 1 *Clearly $\mu_i\sigma_i$ solves $(\nu(P_i), <, L_i)$ if σ_i solves $(\nu(P_i), <, L_i)$. This modified definition of σ_0 gives a more general solution then the definition of σ_0 on page 8. This is necessary to guarantee the completness of the combination algorithm. Consider for example $P_1 = \{x \asymp y\}$, $P_2 = \{u \asymp v\}$. Possible solutions are $\sigma_1(x) = \sigma_1(y) = z$, $\sigma_2(u) = \sigma_2(v) = z$ which gives $\sigma_0(x) = \sigma_0(y) = \sigma_0(u) = \sigma_0(v) = z$. But a more general solution is $\sigma(x) = \sigma(y) = x$, $\sigma(u) = \sigma(v) = u$.*

5.1 $A_1, \dots, A_N, A$ satisfy OUT1

If $\exists(\nu, <, L_1, \dots, L_N) \in \mathcal{L} : A_i(\nu(P_i), <, L_i) \neq$ NO for all $i \in \{1, \dots, N\}$ then return YES. Otherwise return NO.

5.2 $A_1, \dots, A_N, A$ satisfy OUT2

If $\exists(\nu, <, L_1, \dots, L_N) \in \mathcal{L} : A_i(\nu(P_i), <, L_i) \neq$ NO for all $i \in \{1, \dots, N\}$ then return $\sigma_0(\sigma_1, \dots, \sigma_N)^K \nu$ where $\sigma_i = A_i(\nu(P_i), <, L_i)$ and $\sigma_0^K = \sigma_0^{K+1}$. Otherwise return NO.

5.3 $A_1, \dots, A_N, A$ satisfy OUT3

Denote the j-th Element of the output of $A_i(\nu(P_i), <, L_i)$ by $s_{ij}(\nu, <, L_i)$.

For $m = N, N+1, \dots$ do

1. For all $m_1 + \dots + m_N = m$, $m_i \geq 1$, $(\nu, <, L_1, \dots, L_N) \in \mathcal{L}$, return

$$\sigma_0^K(s_{1,m_1}(\nu, <, L_1), \dots, s_{N,m_N}(\nu, <, L_N))\nu,$$

   $\sigma_0^K = \sigma_0^{K+1}$, if $s_{i,m_i}(\nu, <, L_i)$ exists for all $i \in \{1, \dots, N\}$.

2. If there are no $m_1 + \dots + m_N = m$, $m_i \geq 1$, $(\nu, <, L_1, \dots, L_N) \in \mathcal{L}$, such that $s_{i,m_i}(\nu, <, L_i)$ exists for all $i \in \{1, \dots, N\}$ then report the end of the sequence.

5.4 $A_1, \dots, A_N, A$ satisfy OUT4

1. Calculate $S_i = A_i(\nu(P_i), <, L_i)$ for all $i \in \{1, \dots, N\}$.

2. Return $S = \{\sigma_0(\sigma_1, \dots, \sigma_N)^K \nu : \sigma_i \in S_i, \sigma_0^K = \sigma_0^{K+1}\}$.

6 Correctness proof of the combination algorithm

6.1 Completeness

Assume that σ is solution of P, $\sigma\sigma = \sigma$. To prove that the combination algorithm is complete we show the following: There is some $(\nu, <, L_1, \dots, L_N) \in \mathcal{L}$ and there are solutions σ_i of $(\nu(P_i), <, L_i)$; for all solutions τ_i of $(\nu(P_i), <, L_i)$ with $\sigma_i(x) \equiv_{E_i} \lambda_i \tau_i(x)$ for some λ_i, $x \in \text{var}(\nu(P_i))$, we have $\sigma(x) \equiv_E \lambda \sigma_0(\tau_1, \dots, \tau_N)^K \nu(x)$ for all $x \in \text{var}(P)$, with $\sigma_0^K = \sigma_0^{K+1}$ and some substitution λ.

By lemma 1 and lemma 3 there is a solution $\hat{\sigma}$ of $\hat{P}$ with $\hat{\sigma}(x) \equiv_E \sigma(x)$ for all $x \in \text{var}(P)$. Furthermore there are appropriate $\nu, <, L_1, \dots, L_N$ and solutions σ_i of $(\nu(P_i), <, L_i)$, $\hat{\sigma}\sigma_i = \hat{\sigma}$.

Let $\sigma_i(x) \equiv_{E_i} \lambda_i \tau_i(x)$ for $x \in \text{var}(\nu(P_i))$ and let τ_i solve $(\nu(P_i), <, L_i)$. Then by lemma 2 there exists some $K \geq 0$ with $\tau_0^K = \tau_0^{K+1}$, $\tau_0 = \sigma_0(\tau_1, \dots, \tau_N)$. Let μ_i be as in the definition of σ_0 and

$$\hat{\lambda}_i(y) = \begin{cases} \lambda_i(x) & y = \mu_i(x) \\ y & \text{otherwise} \end{cases}$$

Then $\lambda_i(x) = \hat{\lambda}_i\mu_i(x)$. Furthermore let

$$\hat{\lambda}_0(y) = \begin{cases} \hat{\lambda}_i(y) & y = \mu_i(x) \\ y & \mu_i(x) \neq y \quad \forall i \in \{1,\dots,N\}. \end{cases}$$

Then for $x \in \text{var}(\nu(P_i))$, $\mu_i\tau_i(x) \neq x$,

$$\begin{aligned} \hat{\sigma}(x) &= \hat{\sigma}\sigma_i(x) \equiv_{E} \hat{\sigma}\lambda_i\tau_i(x) = \hat{\sigma}\hat{\lambda}_i\mu_i\tau_i(x) \\ &= \hat{\sigma}\hat{\lambda}_0\mu_i\tau_i(x) = \hat{\sigma}\hat{\lambda}_0\tau_0(x) \\ &= \lambda\tau_0(x) \end{aligned}$$

where $\lambda = \hat{\sigma}\hat{\lambda}_0$. For $x \in \text{var}(\nu(\hat{P}))$ we have $\hat{\lambda}_0(x) = x$ and hence $\lambda(x) = \hat{\sigma}\hat{\lambda}_0(x) = \hat{\sigma}(x) = \lambda\tau_0(x)$. Therefore $\sigma(x) \equiv_E \hat{\sigma}(x) = \hat{\sigma}\nu(x) = \lambda\nu(x) = \lambda\tau_0^K\nu(x)$ for all $x \in \text{var}(P)$.

6.2 Correctness

By lemmas 2 and 1 all solutions $\sigma\tau$ produced by the algorithm are solutions of $\hat{P}$ and hence solutions of P.

6.3 Termination

Since $\mathcal{L}$ is finite only the algorithm of section 5.3 needs some observation. If for all $(\nu, <, L_1, \dots, L_N) \in \mathcal{L}$, $m_1 + \cdots + m_N = m$, $m_i \geq 1$, $m = M$, there is some $i \in \{1, \dots, N\}$ such that $S_{i,m_i}(\nu, <, L_i)$ does not exist, then the same is true for all $m > M$. Hence no further solutions exist.

7 Final section

Collecting the results of the previous sections we get

Theorem 1 *Let $E = E_1 \cup \cdots \cup E_N$ be the union of equational theories E_i with disjoint signatures. If for each theory E_i there exists a* restricted unification algorithm *A_i (see page 7) satisfying* OUTj *then we can construct a unification algorithm for E satisfying* OUTj*, too.*

Remark 2 *Assume we have some unification algorithm for theory E_i with additional free constants, i.e. for $E_i' = E_i$, $\Sigma(E_i') = \Sigma(E_i) \cup \{a_1, \dots, a_M\}$. Furthermore the unification algorithm takes as input a set of equations P and constant restrictions $R_x \subseteq \{a_1, \dots, a_M\}$, $x \in \text{var}(P)$, and considers only solutions σ of P such that $a_i \not\ll \sigma(x)$ if $a_i \in R_x$. Then this unification algorithm yields a restricted unification algorithm since the locked variables in L of $(P, <, L)$ can be treated as free constants and $R_x = \{y \in L : y < x\}$ guarantees that the solutions must satisfy $<$.*

If we have some unification algorithm for theory E_i with additional free constants which produces finite and complete sets of unifiers we also have a restricted unification algorithm, since we have to consider only sulutions σ with $a_i \not\ll \hat{\sigma}(x)$ for $a_i \in R_x$ where $\hat{\sigma}$ is a normalization of σ with $\hat{\sigma} \equiv_{E_i} \sigma$ (see page 4 for the definition of normalization).

References

[1] Peter Auer. *Unification with associative functions.* PhD thesis, Technical University Vienna, 1992.

[2] Franz Baader and Klaus Schulz. Unification in the union of disjoint equational theories: combining decision procedures. 1991. To be published in the proceedings of the IWWERT'91.

[3] Alexander Herold. Combination of unification algorithms. In *Proceedings of the 8th Conference on Automated Deduction*, pages 450–496, Springer LNCS 230, 1986.

[4] Claude Kirchner. A new unification method: a generalisation of Martelli-Montanari's algorithm. In *Proceedings of the 7th International Conference on Automated Deduction*, pages 224–247, Springer LNCS 170, May 1984.

[5] Manfred Schmidt-Schauß. Unification in a combination of arbitrary disjoint equational theories. In Claude Kirchner, editor, *Unification*, pages 217–265, Academic Press, 1990.

[6] Jörg H. Siekmann. Unification theory. In Claude Kirchner, editor, *Unification*, pages 1–68, Academic Press, 1990.

[7] Erik Tiden. Unification in combinations of collapse-free theories with disjoint sets of function symbols. In *Proceedings of the 8th Conference on Automated Deduction*, pages 431–449, Springer LNCS 230, 1986.

[8] Kathy Yelick. Combining unification algorithms for confined regular equational theories. In *Proceedings of the 1st International Conference on Rewriting Techniques and Applications*, pages 365–380, Springer LNCS 202, 1985.

ON THE SUBSETS OF RANK TWO IN A FREE MONOID: A FAST DECISION ALGORITHM

(EXTENDED ABSTRACT)

Jean NERAUD

LITP, Institut Blaise Pascal \ LIR

Université de Rouen,

Place Emile Blondel, F-76134 Mont-Saint-Aignan (FRANCE)

Abstract

Given a finite subset X of a free monoid A^*, we define the rank of X as $r(X)=\min\{|Y|: X\subseteq Y^*\}$. The problem we study here, is to decide whether $r(X)\leq 2$ or not. We propose an $O(n.\ln^2 m)$ algorithm, where n stands for the sum of the lengths of the words in X, and m stands for the length of the longest word.

INTRODUCTION

In the context of the free monoids theory, two notions of rank are actually proposed. First, given a set of words X of the free monoid A^*, its rank can be defined as the cardinality $r_1(X)$ of the basis of the free hull of X, i.e. the smallest free submonoid which contains X. The Defect theorem (cf eg [Lo 83]) says that if X is not a code then its cardinality $|X|$ satisfies $r_1(X)\leq|X|-1$. In other words, X is a code iff $r_1(X)=|X|$.

In another way, given a finite subset $X\subseteq A^*$, we can define the rank of X as the smallest cardinality r(X) of a finite set Y satisfying $X\subseteq Y^*$, i.e. such that all the words in X can be written as the concatenation product of words in Y. Clearly, we have the inequality $r_1(X)\leq r(X)$, moreover, this second notion corresponds to the concept of degree, introduced in [HK 86]. This topic meets that of elementariness: a finite subset X is elementary (or independent) iff $r(X)=|X|$. Historically the notion of elementariness applied to morphisms, and its introduction constitued the major step in the delicate proof of the decidability of the famous DOL sequence equivalence problem.

As said in [HK 86], considering the second notion of rank is often more important in a combinatorial point of view. Moreover, although these two concepts seem very close, they leads to different topological properties: for instance, given an elementary set, being maximal is not equivalent to being a maximal code (cf N 88 2]). In [N 90], properties of our rank are established, with regard to some operations on sets.

Within the context of decision problems, an important difference concerns the properties of the integers $r_1(X)$ and $r(X)$. Indeed, it is well known that, given a finite set of words X, deciding whether X is a code can be achieved by applying the classical Sardinas and Patterson algorithm ([SP 50]). This algorithm has been implanted so that it allows to process the set X in time complexity $O(n.\ln|X|)$ (cf [R 82] or [AG 84]), where n stands for the sum of the lengths of the words in X, and $|X|$ stands for the cardinality of X. Moreover, in [S 76] the computation of $r_1(X)$ is given in time $O(|X|.n^2)$.

On the contrary, deciding whether X is elementary, or deciding whether $r(X)$ is smaller than a given integer k, are NP-hard problems (cf [N 88 1]). In this way, it is of interest to examine restrictions of the last problem. For instance, consider a set X whose elements are words with length p, where p is a given positive integer. If $p=2$ then computing $r(X)$ can be done in time $O(|X|^2)$, by determining the family of the connected components of a direct graph associated with X. Howether, for $p\geq3$, deciding whether $r(X)\leq k$ remains NP-complete. As another example, given a finite set of words X, deciding whether $r(X)=1$ can be done by applying the Knuth Morris and Pratt algorithm ([KMP 77]) (indeed we have $r(X)=1$ iff $r_1(X)=1$).

In this paper, we are interested in the following restriction: given a finite set of words X, the problem is to decide whether $r(X)\leq2$ or not. The naive algorithm consists in examining all the two elements subsets $\{x,y\}$ of factors of the words in X, thus we shall solve our problem in time $O(n^5)$. A refinement in $O(n^3)$ can be done by considering restrictive properties of prefixity on $\{x,y\}$. Here, we establish the following:

Theorem: *Given a finite set* $X \subseteq A^*$*, deciding whether* $r(X) \leq 2$ *or not can be achieved in time* $O(n.\ln^2 m)$*, where* n *stands for the sum of the lengths of the words in* X *and* m *stands for the length of the longest word.*

Our proof makes use of a result of [N 88 2], which allows to only consider the biprefix sets $Y=\{x,y\}$ with primitive elements, (i.e. whose elements are only trivial powers of other words). Actually, given a finite set X, we shall conjugate the preceding property with results on the periods ([D 80]) and repetitions ([CR 90]) in a word, for constructing a set S(X), with cardinality $O(\ln^2 m)$, and such that the following holds:

$r(X)=2$ iff there exists a pair of words $(\alpha,\beta) \in S(X)$ with $X \subseteq \{\alpha,\beta\}^*$.

1. PRELIMINARIES

We shall adopt the classical terminology of free monoids (cf e.g. [BP 85]). Given a finite set $X \subseteq A^*$, we set: $n=\sum_{w \in X}|w|$, and $m=\max_{w \in X}|w|$. In an algorithmic point of view, the words and the sets will be represented by linked lists.

1.1. The problem

The problem we study in this paper is a restriction of the general rank problem, *RANK*:

Instance: a finite set of words X, with $r(X)>1$

Question: decide whether or not $r(X)=2$.

(Clearly, we may assume that the basic alphabet, A, satisfies $|A| \geq 3$)

1.2. A property of biprefixity

Let X be a finite subset of A^*. A word $w \neq \varepsilon$ is said primitive iff $w=x^n$ implies $n=1$. According to a result of [N 88 2]:

Proposition 1.1. *Given a finite set* $X \subseteq A^*$*, there exists a biprefix set* $Y \subseteq A^*$*, whose elements are primitive words, such that* $X \subseteq Y^*$*, and such that* $r(X)=|Y|$.

1.3. Overlap of a word

Given a word $w \in A^*$, any word $w' \in \mathit{pref}(w) \cap \mathit{suff}(w) - \{w\}$ is called an *overlap* of w. More precisely, $w' \in A^*$ is an overlap of w iff there exists an unique pair of words (u,v) such that we have $w \in (uv)^* u$, with $u \neq \varepsilon$ and uv a primitive word.

We denote by φ the function which, with every word $w \neq \varepsilon$ associates the longest overlap of w. The computation of $\varphi(w)$ can be done in time linear of $|w|$ by applying the Knuth Morris and Pratt algorithm (KMP algorithm: cf [KMP 77]), which consists in applying the following classical recursive rule:

(1.1) Let $p \in \mathit{pref}(w)$ and let $a \in A$

$$\varphi(pa)= \begin{cases} \varphi(p)a \text{ if this word is a prefix of } w, \text{ and } p\neq\varepsilon \\ \text{else} \begin{cases} \varphi(\varphi(p)a) \text{ if } p\neq\varepsilon \\ \varepsilon \text{ otherwise.} \end{cases} \end{cases}$$

Example. Let $A=\{a,b,c\}$, and $w=(cbcaacbca)^4cbcaacbc$

By iterating the function φ which was defined above, we shall obtain the following overlaps:

-For each of the integers $i \in \{1,..,4\}$, $\varphi^i(w)=(u_1v_1)^{4-i}u_1$, with u_1=cbcaacbc v_1=a.

$-\varphi^5(w)=(u_5v_5)^0u_5$, with u_5=cbc, v_5=aa

$-\varphi^6(w)=(u_6v_6)^0u_6$, with u_6=c, v_6=b,

$-\varphi^7(w)=\varepsilon$.

Moreover, for all $i \in \{1,5\}$, define n_i as the greatest integer such that $(u_iv_i)^{n_i}u_i$ is both prefix and suffix of w. Since $n_i=n_1$ $(2 \le i \le 4)$, the preceding computation leads to introduce the following finite set of tuples:

$\mathit{gener}(w)=\{(u_1,v_1,n_1), (u_5,v_5,n_5), (u_6,v_6,n_6)\}$, with $n_1=4$, $n_5=1$, $n_6=1$.

Lemma 1.2. *Given a word* $w \in A^+$, *there exists a finite set* $\mathit{gener}(w) \subseteq A^+ \times A^* \times (\mathbb{N}-\{0\})$, *such that the following holds:*

1) For each of the overlaps w' *of* w, *there exists an unique tuple* $(u,v,n) \in \mathit{gener}(w)$, *with* uv *a primitive word,* $w' \in (uv)^*u$, *and* n *the greatest integer such* $(uv)^n u$ *is both prefix and suffix of w.*

2) $|\mathit{gener}(w)| \le \log_\Phi |w|$, *where* Φ *stands for the golden section*

3) computing $\mathit{gener}(w)$ *requires time* $O(|w|.\ln|w|)$

Given a word w, we shall construct *gener*(w), by iterating the function φ on w. Claim 2 is established in [D 80], hence claim 3 holds.

Clearly, applying function φ also leads to decide whether a given word w is primitive or not. More precisely, according to [CR 90]:

Lemma 1.3. *Given, a word* $w \in A^+$*, let* *square*(w) *be the set of the primitive prefixes* $x \in A^+$*, with* x^2 *also a prefix of w. Then the following holds:*

1) $|square(w)| \leq \log_{3/2}|w|$

2) the computation of *square*(w) *requires time* $O(|w|.\ln|w|)$

2. THE ALGORITHM

Before to examine the detail of implementation, it is now convenient to explain the main feature of our method. In fact, our algorithm lays upon the following result:

Proposition 2.1. *With the preceding notation, given a finite set of words X, there exists a finite set* $S(X) \subseteq A^* \times A^*$ *such that the following holds:*

1) $|S(X)| \sim O(\ln^2 m)$

2) for all the pairs $(u,v) \in S(X)$*, the set* $\{u,v\}$ *is biprefix*

3) if $r(X)=2$ *then there exists a pair* $(u,v) \in S(X)$*, with* $X \subseteq \{u,v\}^*$

4) computing S(X) *requires time* $O(n.\ln^2 m)$

As a corollary we obtain:

Theorem 2.2. *Given a finite set of words X, with* $r(X) \geq 2$*, deciding whether* $r(X)=2$*, can be implanted so that it runs in time* $O(n.\ln^2 m)$

The scheme of the corresponding algorithm is indicated as follows:

Algorithm

begin

compute the set S(X) of Proposition 2.1; (*/ step 1 /*)

for all the pairs $(u,v)\in S(X)$ **do** (*/ step 2 /*)

begin

construct the deterministic finite automaton $\mathcal{A}$, with behavior $\{u,v\}^*$;

if $\mathcal{A}$ recognizes all the words in X **then** r(X)=2 **else** r(X)>2

end

end

According to Proposition 2.1, claim 4), step 1 requires time $O(n.\ln^2 m)$.

Given the pair of words (u,v), constructing the "flower" automaton $\mathcal{A}$ clearly requires time O(m). Moreover, deciding whether X is included in the behavior of $\mathcal{A}$ requires time O(n). Hence the complete processing of step 2, requires time $O(n.\ln^2 m)$.

3. CONSTRUCTION OF THE SET S(X): THE MAIN STEPS

Let X be a finite set of words. According to Proposition 1.1, if r(X)=2 then there exists a biprefix set Y={x,y}, with x, y primitive, and such that $X\subseteq Y^*$. In fact we are looking for necessary conditions on the words x,y.

3.1. Set *abstract*(X)

The first step is to restraint the search of the words x,y as factors of a set of cardinality 2, *abstract*(X).

Lemma 3.1. *Let* X *be a finite set of words, with* r(X)=2. *then there exists a two element biprefix set, abstract*(X), *such that the following holds:*

1) |*abstract*(X)|=2

1) for all the biprefix sets {x,y}, *if* $X\subseteq\{x,y\}^*$ *then* $abstract(X)\subseteq\{x,y\}^*$

2) constructing abstract(X) *requires time* O(n) (*with* $n=\sum_{w\in X}|w|$)

The computation of *abstract*(X) is done by iterating division of words, as illustrated in the following example:

Example. Let $A=\{a,b,c\}$ and let $Y=\{x,y\}\subseteq A^*$ be a biprefix code. Let $X=\{u_1,u_2,u_3,u_4\}\subseteq Y^*$, with:
$u_1=(ab)^2c^2(ab)^3c$, $u_2=((ab)^2c^2(ab)^3c)^2$, $u_3=(ab)^2c(ab)^2c^2(ab)^3c$, $u_4=((ab)^2c^2)^4$
-Since $r(X)>1$, there exists a two element set $X_0\subseteq X$, with $r(X_0)>1$. Here we have: $X_0=\{u_1,u_3\}$.
-Notice that u_1 is a suffix of u_3. Hence, since Y is a biprefix code, the following set is included in Y^*.

$X_1=\{t_1,t'_1\}$, with $t_1=u_1=(ab)^2c^2(ab)^3c$, $t'_1=u_3u_1^{-1}=(ab)^2c$

-In a similar way, the following set is incuded in Y^*:

$X_2=\{t_2,t'_2\}$, with $t_2=t'_1{}^{-1}t_1=c(ab)^3c$, $t'_2=t_2=(ab)^2c$

-By iterating the preceding method, we shall construct a sequence of two elements sets (X_k), with $X_k\subseteq\{x,y\}^*$. Since the sum of the lengths strictly decreases, there exists an integer p, with X_p a biprefix code. We set *abstract*$(X)=X_p$. Here we have:

abstract$(X)=X_3=\{t_3,t'_3\}$, with $t_3=t_2t'_2{}^{-1}=cab$, $t'_3=(ab)^2c$. Since the words have been represented by linked lists, computing the set X_k $(1\le k\le 3)$ requires time linear of $n_{k-1}-n_k$, with $n_k=\sum_{w\in X_k}|w|$. Consequently, computing the set *abstract*(X) requires time $O(n)$, with $n=\sum_{w\in X}|w|$.

3.2. The three main cases

Let *abstract*$=\{w_1,w_2\}$. Since $\{w_1,w_2\}\subseteq\{x,y\}^*$, one of the following cases occur:

a) one of the words w_1,w_2 belongs to $(xyY^*\cap Y^*yx)\cup(yxY^*\cup Y^*xy)$
b) one of the words w_1,w_2 belongs to $t^2Y^*\cup Y^*t^2$, with $t\in\{x,y\}$
c) $\{w_1,w_2\}\subseteq(xyY^*xy)\cup(yxY^*yx)$

In each case, we look for new necessary conditions on the words x,y.

3.3. The case where one of the words w_1,w_2 belongs to $xyY^*\cap Y^*yx$

Assume that $w_1\in xyY^*\cap Y^*yx$. Clearly, the word x is an overlap of w_1:

(3.1) there exists a tuple $(u,v,p)\in$ *egener*(w_1) such that $x\in(uv)^*u$

A combinatorial study leads to the following result:

Lemma 3.2. *With condition (3.1), there exists a set* I *of factors of* w_1, *with* $|I|\le 5$, *such that one of the following conditions holds:*

(i) There exist an integer $i\in[1,p-2]$, *and an integer* $k\ge 1$, *such that:*

$$x=(uv)^i u,\ y^k\in(vu)^* v$$

(ii) $x\in I$

Moreover, the set I *and the integer* k *are computable in time linear of* $m=\max\{|w|:w\in X\}$.

In an algorithmic point of view, Lemma 3.2. leads to compute two sets, P_1,Q_1, whose elements are respectively the pairs $(u,v)\in A^*\times A^*$, and the words $x\in I$, for all the tuple $(u,v,p)\in gener(w_1)\cup gener(w_2)$.

According to Lemma 1.2, the computation of the sets P_1, Q_1 will be done in time $O(m.\ln m)$, moreover, we have $|P_1|,|Q_1|\sim O(\ln m)$.

3.4. The case where one of the words w_1,w_2 belongs to x^2Y^*

With the hypothesis of the preceding section, let $Y=\{x,y\}$ and let $w_1\in abstract(X)$ such that $w_1\in x^2Y^*$. By definition, the word x belongs to $square(w_1)$. This leads to compute the set $Q_2=square(w_1)\cup square(w_2)$. According to Lemma 1.3, this computation will be done in time $O(m.\ln m)$. Moreover, we have $|Q_2|\sim O(\ln m)$.

Clearly, the case where a word w_1 belongs to Y^*x^2 will be examined in a similar way, and leads to compute a set Q'_2.

3.5. The case where $\{w_1,w_2\}\subseteq(xy.Y^*\cap Y^*.xy)\cup(yx.Y^*\cap Y^*.yx)$

Without loss of generality, we may assume that xy is a overlap of w:

(3.2) there exists a tuple $(u,v,p)\in gener(w_1)$ such that $xy\in(uv)^*u$.

As in Section 3.3, a combinatorial study leads to the following result:

Lemma 3.3. *With condition (3.2), there exists a finite set* I *of factors of* w_1, *with* $|I|\le 4$, *such that one of the following conditions holds:*

(i) There exist an integer $i\in[2,p-2]$, *and an integer* $k\ge 1$, *such that:*

$$xy=(uv)^i u,\ y^k=v$$

(ii) $xy\in I$.

Moreover, the set I *and the integer* k *are computable in time linear of* m.

This leads to compute, in time $O(m.\ln m)$ two sets, P_3, R_3, whose elements are respectively the pairs $(u,v) \in A^* \times A^*$, and the words $xy \in I$, for all the tuple $(u,v,p) \in gener(w_1) \cup gener(w_2)$. According to Lemma 1.2, we have $|P_3|, |R_3| \sim O(\ln m)$.

3.6. The ultimate condition

Let $P = P_1 \cup P_3$, let $Q = Q_1 \cup Q_2 \cup Q'_2$, and let $R = R_3$. As a consequence of the preceding results, if $r(X)=2$:

(3.3) There exists a biprefix set $\{x,y\}$, with primitive elements, which satisfies one of the three following conditions:

1) there exists a pair $(u,v) \in P$ such that $x \in (uv)^* u$, and $y^k \in (vu)^* v$ $(k \geq 1)$

2) $x \in Q$

3) $xy \in R$

Clearly, the computation of P,Q,R have been done in time $O(m.\ln m)$, moreover: $|P|, |Q|, |R| \sim O(\ln m)$.

A new combinatorial study leads to establish the ultimate necessary condition:

Lemma 3.4. *With condition (3.3) there exists a set* $S \subseteq A^* \times A^*$ *which satifies the following properties:*

1) for all the pairs $(u,v) \in S$, *the set* $\{u,v\}$ *is biprefix*

2) there exists a pair $(u,v) \in S$ *such that* $x, y \in \{u,v\}^*$

3) $|S| \sim O(\ln m)$

4) computing S *requires time* $O(m.\ln m)$

Let S(X) be the union of the sets S thus constructed for all the pairs $(u,v) \in P$, and all the words in $Q \cup R$. Clearly the set S(X) satisfies the required conditions of Proposition 2.1.

REFERENCES

[AG 84] Apostolico A. and R. Giancarlo. Pattern matching implementation of a fast test for unique decipherability, Information Processing Letters 18 (1984) 155-158.

[BP 85] Berstel J. and D. Perrin ."Theory of codes", Academic Press, 1985

[CR 90] Crochemore M. and W. Rytter. Manuscript (1990)

[D 80] Duval J.P. Contribution à la combinatoire du monoïde libre. Thèse d'Etat, Université de Rouen (1979)

[HK 86] Harju T. and J. Karhumäki. On the Defect theorem and simplifiability, Semigroup Forum Vol.33 (1986) 199-217

[KMP 77] Knuth D., Morris J. and V. Pratt. Fast pattern matching in strings. SIAM J. Comput. 6 (1977) 323-350.

[Lo 83] Lothaire M. "Combinatorics on the words", Encyclopedia of mathematics and its applications, Addison-westley Publishing Company, 1983.

[N 88 1] Néraud J. Elementariness of a finite set of words is co-Np-complete, Theoret. Info. and Appl. Vol.24, N°5 (1990) 459-470

[N 88 2] Néraud J. On the deficit of a finite set of words, Semigroup Forum Vol 41-1 (1990), 1-21.

[N 90] Néraud J. On the rank of the subsets of a free monoid, to appear in TCS, Vol 102 (1993).

[R 82] Rodeh M. A fast test for unique decipherability based on suffix trees, IEEE Trans. Information Theory IT-28 (1982) 648-651.

[S 76] Spehner J.C. "Quelques problèmes d'extension, de conjugaison, et de présentation des sous-monoïdes du monoïde libre", Thèse de Doctorat d'Etat, Université de Paris VII (France), 1976.

[SP 50] Sardinas A. and G.W. Patterson. A necessary and sufficient condition for the unique decomposition of coded messages. Reshearch Division Report 50-27. Moore School of EL.Eng., University of Pennsylvania, 1950.

A SOLUTION OF THE COMPLEMENT PROBLEM IN ASSOCIATIVE-COMMUTATIVE THEORIES

KOUNALIS Emmanuel* LUGIEZ Denis POTTIER Loic*****
LIR INRIA INRIA

ABSTRACT: We show in this paper that the problem of checking whether there are ground instances of a term **t** which are not instances of the terms **t1**, ..., **tn** modulo sets of associativity and commutativity axioms is decidable. This question belongs to the the well-known class of *complement problems*.

Its solution provides a formal basis for automating the process of learning from examples, verifying the sufficient-completeness property of algebraic specifications, designing associative-commutative compilation algorithms, finding solutions of systems of equations and disequations in associative and commutative theories, etc.

KEYWORDS: The Subsumption Lattice of First-Order terms, Associative and Commutative Reasoning, AC-Disunification, Inductive Learning, Pattern-Matching, Sufficient-Completeness.

*) Laboratoire d'Informatique de Rouen , BP 118 Place Emile Blondel B.P. 118 76134 Mont-Saint-Aignant,- FRANCE.

**) Campus scientifique - BP. 239, 54506 Vandoeuvre-lès-Nancy cedex, FRANCE, {kounalis, lugiez}@loria.crin.fr

***) 2004 route des Lucioles, Sophia Antipolis, 06565 ValbonneCedex, FRANCE, pottier@mirsa.inria.fr

1. INTRODUCTION

The need to be able to reason about (first-order) *terms* with associative and commutative functions is fundamental in many Computer Science applications including logic and functional programming, automated reasoning, specification analysis, machine learning, Computer Algebra etc... Associativity and commutativity are typical axioms that are more naturally viewed as *structural* axioms (defining a congruence relation on terms) rather than *simplifiers* (defining a reduction relation).

We denote **variables** by x, y, z... and denote **function symbols** by f, g, h,... Each function symbol has an associated arity. Function symbols with arity 0 (zero) are called **constant symbols** and denoted by a, b, c, c_1, c_2,.... A **(first-order) term** is a variable, a constant or of the form f(t_1, ..., t_n), where f is a function symbol with arity n and t_1, ..., t_n are terms. The symbols t, s, r, t_1, ... denote terms. Let **T(F,X)** denote a set of terms built out of function symbols taken from a finite vocabulary (signature) **F** and a denumerable set **X** of variables. We assume that **F** contains at least one constant symbol. Thus, the set of **ground terms** (the **Herbrand Universe**), i.e., terms containing no variable, is non-empty. An equation over T(F,X) is an element of **T(F,X)** X **T(F,X)**. The equation (t,s) will be written as t = s. Given a set **E** of equations, the **equational theory of E** is the set of equations derivable from **E** by a finite proof, using reflexivity, symmetry, transitivity, replacement of equals, and instantiation as inference rules over equations. A function symbol f in **F** is **commutative** if and only if it satisfies the equation of the form **f(x,y) = f(y,x)**, where x, y are distinct variables. A function symbol f in **F** is **associative,** if and only if, it satisfies the equation of the form **f(f(x,y),z) = f(x,f(y,z))**, where x, y, z are distinct variables. We often refer to a function symbol that is both **associative** and **commutative** as an **AC-function.** Two terms t and s are said to be **associative-commutative equivalent**, written as $t =_{AC} s$, if and only if they are equivalent under the equational theory of the associative and commutative axioms (**AC-theory**).

A **substitution,** $\sigma = \{x_1 \leftarrow t_1, ..., x_n \leftarrow t_n\}$ where $x_i \neq t_i$, assigns terms to variables. The term, tσ, which is the result of applying the substitution σ to a term t, is said to be an **instance** of t. tσ is obtained by simultaneously replacing all occurrences of each variable x_i in t by t_i. If each term t_i in σ is ground, then tσ is said to be a **ground instance** of t. If s and t are terms and AC is an associative-commutative theory, then s is an **AC-instance** of t, if there is an substitution σ such that $s =_{AC} t\sigma$.

In this paper we deal with the following Complement Problem in Associative-Commutative theories (for a short **CPAC-problem**):

INSTANCE: A finite set t, t_1, ..., t_n of first-order terms over T(F,X) with AC-functions.
QUESTION: Does there exist ground terms that are AC-instances of t but not AC-instances of any of the terms t_1, ..., t_n ?

Motivation:

Generalization is an important operation for programs that **learn from examples (inductive learning)**. A formal model for inductive inference has been proposed by Plotkin [PL,71], Angluin and Smith [AS, 83]. A formula of the form **G**:= t/ t_1,..., t_n is a **generalization** iff the set of ground instances of t which are not instances of t_1,..., t_n is non-empty. Recently Lassez and Marriot [LM,87] solved the CPAC-problem in the case where no AC-functions are involved in the terms t, t_1,..., t_n. However, in the learning from examples paradigm most of the symbols appearing in the set of examples and counter examples are AC-functions (notarious examples are the connectives *and, or* etc. (see Michalski [MICH, 83]). For instance, if a set of concept examples leads to the term **or(x,y)** and a set of counter examples is **or(x,x)**, **or(f(z),c)**, **or(f(y1), f(y2))**, then from immediate inspection it is not obvious that the formula **or(x,y) / or(x,x)**, **or(f(z),c)**, **or(f(y1),f(y2))**, provides no concept to be learned. Now, if in the **CPAC**-problem the term t is the generalization of a set of examples and the terms t1, t2,..., tn are counter examples, then its solution allows to check whether the formula **G**:= t/ t1, t2,..., tn is a generalization.

Functional languages allow to define functions by a set of rules such as: **0 + y → y, x + succ(y) → s(x+y)** together with a priority on this set of rules to deal with ambiguous patterns (i.e., a set of left-hand sides). However, for efficiency reasons the above definition should be compiled in a piece of code which is more suited to machine: (x+y) = **if** x = 0 **then** y **elsif** y=s(y') **then** let y=s(y') in s(x+y') **else no_match.** Nowdays, many algorithms for **compiling pattern-matching** definitions exist (see Peyton-Jones [P-J,87], Schnoebelen [SCH,88], Kounalis-Lugiez [KL,91]). Unfortunately, these algorithms apply either to a set of patterns without AC-functions, or to linear patterns. However, if in the **CPAC**-problem the term t is f(x1,x2,...,xk) and the terms t1, t2,..., tn are the patterns of the k-ary function f, then its solution will allow us to design effective algorithms to compile pattern-matching definitions.

The sufficient-completeness property of (equational) specifications of **Abstract Data Types** allows to ensure that every data item can be constructed using only constants and operations of a signature (*no junk*). This property has been investigated for specifications under certain conditions (Huet and Hullot[HH,82], Kounalis [KOU, 85],etc). Unfortunately, these algorithms cannot deal with specifications which include AC-axioms. However, if in the **CPAC**-problem the term t is f(x1,x2,...,xk), where f is the function to be checked for sufficient-completeness, and the terms t1, t2,..., tn are the left-hand sides of an equational specification, then its solution will allow us to design algorithms to verify the sufficient completeness property of convergent specifications with AC-functions.

Systems of equations and disequations $\{l_i = r_i: 1 \leq i \leq n,\ d_j = g_j: 1 \leq j \leq n\}_{AC}$ have many applications to Artificial Intelligence. The solution of such a system is a set of substitutions of the variables occurring in l_i, r_i, d_j, g_j, such that the instantiated terms of equations become equal, and the instantiated terms of disequations become inequal with respect to a given AC-theory. Colmerauer [COL,84] has discussed this problem in the framework of **Logic Programming** and given an algorithm to solve this problem in the case of an empty theory (see also Lassez and Marriott [LM,87] , Maher [MAH,88]). However, if we assume that some of the functions appearing in the system of equations and disequations are AC-functions then their results do not hold anymore.Consider, for instance, the system S = $\{\mathbf{x = f(a,y)),\ y \neq x}\}_{AC}$.The problem here consists of finding substitutions of the variables x and y such that the system S has a solution. However, if in the **CPAC-** problem the term t is P(f(a,y), y)) and the set of terms t1, t2,..., tn consists only of the term P(z,z), then its solution will allow us to find a complete set of solutions of a system S of equations and disequations. Note that there is an easy algorithm to transform S into the set t, t1, t2,..., tn.

How to solve the Complement problem (CPAC) ?

The main tool to solve the **CPAC**-problem is the concept of *pattern trees* which are trees whose nodes are labeled by flattened terms. However, when the signature contains several constant symbols and only one function symbol of arity greater than zero, then the solution of the **CPAC**-problem is reduced to verify the emptyness of differences of particular subsets of N^m using properties of additive semi-groups of N. In both cases, the correctness proof is non-trivial and requires the use of sequences of terms with complex combinatorial properties. At first glance this result may appear surprising since closely related problems are undecidable (see Kapur, Narendran, and Zhang [KNZ, 86], Treinen [TRE,90]).

Layout of the paper:

The structure of this paper is as follows: In Section 2 we present an overview of the proposed methods on two examples. In Section 3 we summarize the additional basic material which is relevant to this work. In Section 4 we provide the decision procedures for the **CPAC**-problem. In particular, we first show how to solve the **CPAC**-problem in the case where the signature contains several constant symbols and at least two function symbols of arity greater than zero, and next consider signatures that contain several constant symbols and only one function symbol of arity greater than zero.

2. OVERVIEW OF THE SOLUTION: TWO EXAMPLES

Before discussing the technical details of how to check whether there exist ground instances of a term t which are not AC-instances of a set $S = \{ t_1, ..., t_n \}$, we first describe how these methods work on two simple examples. Each of the method solely depends on the structure of the signatures to be used : *Basic* signatures are those containing several constant symbols and only one function symbol of arity greater than zero. *Extended* signatures are those containing several constant symbols and at least two function symbols of arity greater than zero

EXAMPLE A: Consider, for instance, the term t = f(x,y) and the set S of terms { t_1= f(x1,x1), t_2=f(x1, f(x1,y)), t_3 =f(x1, g(y1)) }, over T({f,g,0}, X} where f is an AC-function symbol, g is a function of arity 1, and 0 is a constant symbol. Here the signature is extended.

The first step of our method consists of considering terms in a *flattened* version (see preliminaries definitions below) i.e., in a form in which all nested occurrences of associative-commutative function symbols are stripped, and where the order of arguments of such operators is not significant. For example, the flattened version of the term t2 is the term f(x1, x1, y).

Having computed the flattened version of the terms t, t_1, ..., t_n our second step consists of constructing a *complete pattern tree* for the flattened term t (see definition 4). By a *pattern tree* of t (see definition1) we mean a finitely branching tree whose root is labeled by the term t and such that an internal node labeled by s has successors all possible different flattened terms (modulo variable renaming) derived from s, i.e., obtained by replacing some variable in s by some term of the form g(x1, ..., xn), where g is a function symbol of arity n (in the signature F) and x1, ..., xn are distinct variables not appearing in s. For instance the following pattern tree is suitable for checking the **(CPAC)**-problem:

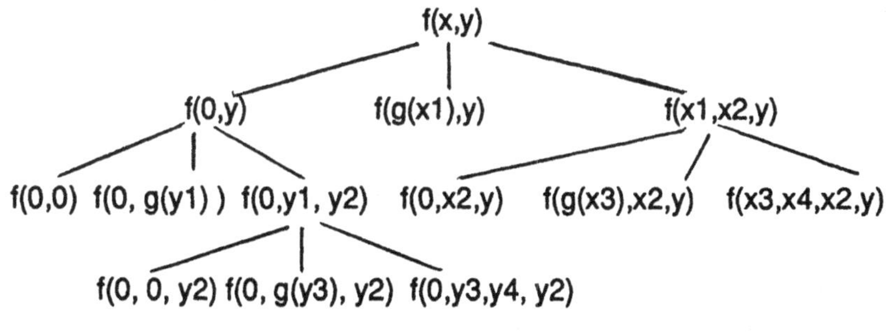

The key idea undrlying the construction of such a pattern tree **T** consists of extending variables in node labels whenever these nodes are AC-unifiable with some term t_j in S. In general this process is infinite and therefore we need some way to stop it: a node label s in a pattern tree is a leaf if

1) s is a term which is an AC-instance of some t_j in S (e.g., the terms f(g(x1),y) , f(0,0), f(0, g(y1)), f(0, 0, y2), f(0, g(y3), y2), f(g(x3),x2,y).

2) s is a term which is not AC-unifiable with some term in S (In the example here each leaf label is AC-unifiable with some term in S).

3) s is a term whose any non-variable subterm s/u has either no variable which corresponds to a function symbol or to a non linear variable in some t_j in S or a number of arguments greater than the maximal number of arguments of an AC-function in S (e.g., the terms f(0,y3,y4, y2), f(x3,x4,x2,y)).

The terms as partitioned above are said to be of *type* 1, 2, and 3 (see definition 4). A pattern tree whose all leaf labels are of type 1, 2, 3 is said to be *complete* (see definition 4).

Having computed a complete pattern tree our next step consists of testing whether there exist leaf labels in **T** of type 3. The reason for that is the following: if all leaf labels in T is of type 1 or 2, then the (**CPAC**)-problem is resolved since terms of type 1 have all ground instances AC-equivalent to the ground instances of terms in S, whereas terms of type 2 have no ground instance AC-equivalent to the ground instances of terms in S. Unfortunately, the situation with terms of type 3 is more complicated. Terms of type 3 are further partitioned into two sets according to the number of arguments of AC-function symbols occurring in them (see definition 6). In the present example all terms of type 3 are of type 3b (see definition 6) since the number of arguments of the function symbol f in the terms f(0,y3,y4, y2), f(x3,x4,x2,y)) is greater than 3: the maximal number of arguments of an AC-function in S. We must note here that the existence of terms of type 3a ensures that (**CPAC**)-problem is resolved since terms of type 3a have ground instances which are not AC-equivalent to any ground instance of terms in S (see theorem 4). However, in the absence of terms of type 3a we have to check whether the terms of type 3b contain ground instances which are not AC-equivalent to the ground instances of terms in S.

This consists of the last step of the method. To check terms of type 3b we compute an *answer set for* S (see definition 7) which in our case is the set A(S) = { 0, f(0,0), f(0,g(x)), g(0), g(f(x,y)), f(0,0,0), f(0,0,g(x)), f(g(z),0,g(x)), f(g(z),g(w),g(x)), f(0,0,0,y), f(0,0,g(x),y), f(g(z),0,g(x),y), f(g(z),g(w),g(x), y). We then verify whether all instances of a term of type 3b which are obtained by using terms in A(S) are AC-instances of S. Here, we can easily verify that all instances of the terms f(0,y3,y4, y2), f(x3,x4,x2,y)) by using elements of A(S) are AC-instances of the terms in S and this resolves the (**CPAC**)-problem (see theorem 5): *each ground instance of the term* t = f(x,y) *is an* AC*-ground instance of the terms* t_1= f(x1,x1), t_2=f(x1, f(x1,y)), t_3 =f(x1, g(y1)) }.

EXAMPLE B: Consider now the following example of the **CPAC**-problem. t = +(x,x,y,y) and t_1= +(x1,x1,x1, y)) over T({+,a}, X}, where + is an AC-function symbol, and a is a constant symbol. Therefore the signature here is basic.

In general, to tackle the **CPAC**-problem in the case where the signature is basic, we first consider terms as sums of variables, where + is the usual AC-function symbol. To denote such terms we shall use the notation $\Sigma_i a_i x_i + \Sigma_j a_j C_j$ which means that the variable x_i appears a_i times and the constant C_j appears a_j times in a term. In other words to any term of the **CPAC**-problem we associate vectors $(a_1, a_2, ... a_k)$ of N^k. We then reduce the search of solutions of the **CPAC**-problem in a finite subset of N^k (**a test domain**).

For instance, in our case , let t be the term $2z_1+2z_2$ and t_1 be the term $3\ x_1+ x_2$. Since the number of constant symbols in the signature is one , by applying theorem 6, we get that the test domain is the segment [0;8]. This means that we must verify whether there exists an instance of t obtained by replacing the variables in t by elements from the finite domain n*a, where n varies from 1 to 8 and a = (2,2) , which is not an AC-instance of t_1. However, all these instances of t are instances of t_1, we may conclude by using theorem 6 *that every ground instance of* t *is an AC-instance of* t_1.

3. PRELIMINARIES

In order that this paper be self-contained, this section contains an outline of further definitions (besides the ones presented in the introduction of this paper) and results related to AC-terms.

Let AC be an associative-commutative theory. By $[t]_{AC}$ (or simply [t]) we denote the AC-equivalence class containing a term t in T(F,X). If s and t are terms and AC is an associative-commutative theory, then: s is an AC-instance of t iff there exists s' in [s] such that s' is an instance of t. Further, suppose s and t are terms in an AC-theory. Then s and t are said to be **AC-unifiable** if there is a substitution $\bar{s}$ such that $s\bar{s} =_{AC} t\bar{s}$. In this case we say that s is an **AC-unifier** of s and t.

A term that involves associative -commutative function symbols is represented in **flattened** form, that is, no argument to an AC-function f is a term whose outermost symbol is f itself. For example, if f is AC, then f(a,f(b,c)) is represented as f(a,b,c). (In other words f is treated as a *varyadic* symbol). Flattening a term with respect to a function f can be done as follows: *firstwe* represent the term in the right-associative form. Such a term will be of the form $f(t_1, f(t_2, ..., f(t_{n-1}, t_n)...)$ where $t_1, ..., t_n$ do not start with f. *Then* we simply represent the term as $f(t_1, ..., t_n)$. If t is a term, then **Flat(t)** denotes the flattened form of t. In the following we shall make free use of the following lemma:

Lemma: *Suppose s and t are terms. Then $s =_{AC} t$ iff $Flat(s) =_P Flat(t)$, and $s =_{AC} t\bar{s}$ iff $Flat(s) =_P Flat(Flat(t)Flat(\bar{s}))$, where $=_P$ is the permutative congruence on subterms of AC-functions.*

Given a term t, **arg(t)** denotes the set of arguments of t, i.e., if $t = f(t_1, ..., t_n)$, then **arg(t)**= $\{t_1, ..., t_n\}$. Further , **|arg(t)|** denotes the cardinality of arg(t). Also the **depth(t)** of a term t is defined to be the depth of the maximal path in the tree representation of t.

If t is a term, then **dom(t)** denotes the set of positions (occurrences) in a term t and the expresion t/u denotes the **subterm of t at position u.** Also **t(u)** denotes the symbol of t at position u. A position u in a term t is a **variable position** if t(u) is a variable. Further, a position u in t is said to be a **non-linear variable position** if t(u) is a variable position and the variable x=t(u) appears more than once in t. Let **sdom (t)** denote the set of function symbols or non-linear variable positions in t and **V(t)** to denote the set of all variables that occur in t. A term t is **linear** iff no variable in **V(t)** appears more than once in t.

In the second part of the next section we need another representation of flattened terms. Let + be an AC-function. A flattened term $t = +(s_1,...,s_k)$ where the s_i are either variables or constants will be written as $\Sigma_i a_i s_i$, where a_i is the number of occurrences of s_i in t. Some a_i can be zero, but the sum of a_i cannot be zero. For instance the term t = +(x,a,b,x,a) is written as 2x+2a+ b. Let N be the the set of non-negative integers. By [a; b] we denote a **closed segment** of N. Also by aN we denote the set of non-negative multiples of a. Let $e_1,...,e_n$ be a canonical basis of N^n. The ith **coordinate** of a vector x in N^n is denoted by x_i. Further, let $a_1,...,a_n$ be positive integers. Then we denote by (a) the **additive semigroup** of N generated by the a_i's. Also gcd(a) denotes their **greatest common divisor,** and lcm(a) their **least common multiple**. The following result is well-known from number theory (**Frobenious problem**):

Lemma: *There exists a least integer c(a) such that every integer c(a) + k*gcd(a) is in N, where k∈N.* *c(a) is said to be the* ***conductor*** *of (a).*

4. DECISION PROCEDURES

In this section we set the machinery to resolve the **CPAC**-problem stated in the introductory remarks of this paper, i.e., the problem of whether, in a sequence t, t1,...,tn of terms, there exist ground instances of t which are not instances of any t1,...,tn modulo the associativity and commutativity axioms.

Clearly, the major problem in solving the **CPAC**-problem is the unbounded number of ground substitutions of t one has to verify for instances of the terms t1,...,tn. To motivate our efforts in removing this problem we first consider the case of signatures with several constants and at least two function symbols of arity greater than zero **(extended signatures)**, and next consider the case of signatures with several constants and only one function symbol of arity greater than zero **(basic signatures)**. The reason of such a partitition is that the extended signatures allows the construction of flattened terms of any depth. Note that this is not the case of basic signatures.

4.1 (CPAC)-PROBLEM FOR EXTENDED SIGNATURES:

Let us first consider the case of signatures with several constants and at least two function symbols of arity greater than zero and several constant symbols(extended signatures). The key to our approach in this case is the construction of *Pattern trees* :

Definition 1: Suppose t is a flattened term. Then the **sons** of t at a fixed variable x are all possible different flattened terms (modulo variable renaming) *derived* from t i.e., obtained by replacing the variable x in t by some term of the form g(x1, ..., xn), where g is a function symbol of arity n (in a signature F) and x1, ..., xn are distinct variables not appearing in s.

Example 1: Suppose that F={a, f, h} with f to be an AC-function symbol. Then the sons of f(x,y,z,y) are the terms: {f(x,a,z,a), f(x, x_1,x_2, z, x_1,x_2) f(x,

$h(x_1),z, h(x_1))$ }. These terms are obtained by replacing the variable y in $f(x,y,z,y)$ by the terms a, $h(x_1)$, and $f(x_1,x_2)$.

Definition 2: A **Pattern tree T** of a flattened term t is a labeled tree **T** whose root is labeled by t, and such that an internal node labeled by a flattened term s has successors the sons of s.

Example 2: Let F={a, f, h} and let f be an AC-function symbol, then a pattern tree of $f(x,y,z,y)$ is:

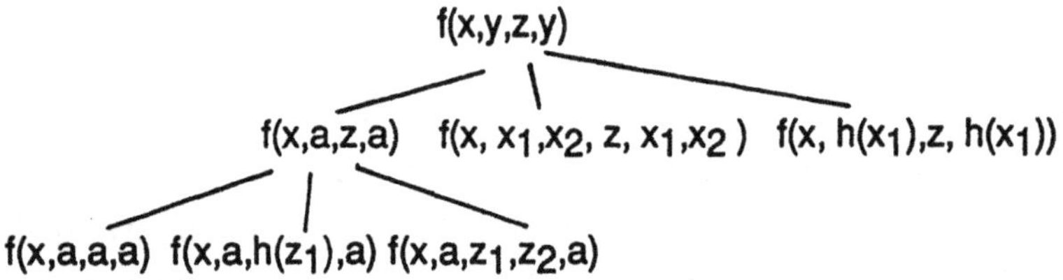

It follows from the above definition that in a pattern tree of a flattened term t, (i) the label of any node is an instance of t,(ii) the set of ground instances of t is equal to the set of ground instances of all leaf labels, and (iii) for any ground substitution η there exists a leaf label r and a ground substitution σ, such that $t\eta= r\sigma$.

As we already have stated in the introductory example of this paper, the method of solving the **CPAC-problem** for extended signatures is based on the construction of a suitable finite pattern tree **T** of a flattened term t. In trying to compute such a tree there are several questions that come naturally in mind:

Q1) *Which node labels must be expanded* ?

Q2) *Which variables in them must be instantiated*?

Q3) *When the construction halts*?

Let us now set the machinery to resolve these queries. The following definition gives the answer of the first two queries:

Definition 3: Suppose t is a flattened term and S= $\{t_1, ..., t_n\}$ is a set of flattened terms. Let k be the maximal number of arguments of an AC-function in S= $\{t_1, ..., t_n\}$. t is said to be **extensible** with respect to S=$\{t_1, ..., t_n\}$ if there exists a term s in $[t]_p$, a variable position u.i in s, and an integer $j \leq n$ such that

u.i is in **sdom**(t_j) [i.e, u.i is a function symbol or a non-linear variable position in t_j];

and $|arg(s/u))| \leq k$, [i.e, the number of arguments of s/u is less or equal than maximal number of arguments of an AC-function in S= $\{t_1, ..., t_n\}$],

or there exists v.j in **sdom**(s) such that s(u.i) = s(v.j) and s(u) ≠ s(v) [i.e, s(u.i) = s(v.j) = x (x is a non-linear variable) and different fathers)

Otherwise, t is said to **cover** S = $\{t_1, ..., t_n\}$.

Example 3: Let F={a, f, h} and let f be an AC-function. The term t= f(a, x_1, z_1, h(x_2)) is extensible with respect to the flattened term t_j = f(x,y,z,y) since the variable x_1 is in the position 1 in the term s (x_1,a, z_1, h(x_2)) $=_P$ t, and 1 is a non-linear variable position in t_j. However, the term t= f(h(x_1),a, h(z_1), h(x_2)) is not extensible with respect to t_j since no term in $[t]_P$ contains a variable position which belongs to sdom(t_j). The term t= f(x, x_1, x_2, z, x_3, x_4) covers t_j since |arg(f(x, x_1, x_2, z, x_3,x_4))| > |arg(f(x,y,z,y))|.

It follows from the above definition that if a non-ground term t covers a set S = { t_1,...,t_n} of flattened terms, then for all $j \leq n$, and for every variable x in $[t]_P$ ***either*** x is in position that corresponds to a function symbol or to a non-linear variable in t_j ***or*** x is an argument of an AC-function subterm in $[t]_P$ the number of arguments of which is greater than k.

Morever, definition 3 provides a way to compute the pattern trees we are interested in: these pattern trees are obtained by deriving node labels which are extensible with respect to given set { t_1,...,t_n} of flattened terms. Let us now resolve the last query stated above. To do it we need the following definition :

Definition 4: Given a set {t_1, ..., t_n} of flattened terms and a flattened term t, let **T** be a pattern tree of t. A node label s in **T** is said to be of

type 1, if s is an AC-instance of some t_j,

type 2, if s is not AC-unifiable with t_j, for any $j \leq n$,

type 3, if s is neither of type 1 nor of type 2 and covers {t_{i1}, ..., t_{ik}}, where { i1, ..., ik } is a subset of {1,...,n} and s is AC-unifiable with t_{ij}, for any $j \leq k$.

A pattern tree **T** of t is said to be **complete** if each node label of type 1, 2, 3 is a leaf.

Example 4: Let F={ a, b, f, h} and let f be an AC-function. Let t = f(x,y), t_1 = f(x,y,y) , t_2 = f(x,a) , t_3 = f(b,b). Term s=f(x,y,z) is not of type 1,2,3 since s is AC-unifiable with the term t_1 and does not cover it. The term f(a,x) is of type1 since it is an AC-instance of t_2. Term f(h(x),b) is of type 2 since is not AC-unifiable with S= $\{t_1,t_2,t_3\}$. Term s=f(h(x),h(y), h(z) is of type 3 since s is AC-unifiable with the term t_1 and covers it.

Given a flattened term t and a set S= $\{t_1, ..., t_n\}$ of flattened terms, there always exists a complete pattern tree T for t which is finite (see theorem 1). The reason for that is that the number of terms s whose depth is bounded by d (the depth of S) and such that no subterm of s has more than k +1 arguments is finite.

Theorem 1: *If S= $\{t_1, ..., t_n\}$ is a set of terms, a complete pattern tree T of a term t is finite.*

Let us now show how complete pattern trees allow to solve the **CPAC**-problem. This will be done by studing the fundamental properties of leaf labels of a complete pattern tree.

Definition 5: Let S= $\{t_1, ..., t_n\}$ be a set of flattened terms, and let **T** be a complete pattern tree of a flattenned term t. A leaf label s in T is a **quasi-AC-instance** of S if for every ground instance sh of s there exists a term t_j in S such that sh is an AC-instance of t_j.

The previous definition allows us to solve the **CPAC**-problem by checking whether a leaf label of a complete pattern tree of a flattened term t is a quasi-AC-instance of a set S = $\{t_1, ..., t_n\}$ of flattened terms: **each ground instance of t is AC-equivalent to ground instances of S iff each leaf label of a complete pattern tree of t is a quasi-AC-instance of a set S = $\{t_1, ..., t_n\}$.**

The following theorem deals with leaf labels of type 1:

Theorem 2: *Let S = $\{t_1, ..., t_n\}$ be a set of terms, and let T be a complete pattern tree of a flattened term t. Leaf labels in T of type 1 are quasi-AC-instances of S= $\{t_1, ..., t_n\}$.*

Leaf labels of type 2 enjoy the following property:

Theorem 3: *Let $S = \{t_1, ..., t_n\}$ be a set of terms, and let T be a complete pattern tree of a flattened term t . Leaf labels in T of type 2 are not quasi-AC-instances of $S= \{t_1, ..., t_n\}$.*

The situation with terms of type 3 is more complicated.To proceed with, we first need the following:

Definition 6: Let $\{t_1, ..., t_n\}$ be a set of flattened terms, and let **T** be a complete pattern tree of a flattenned term t. Assume that k is the maximal number of arguments of an AC-function in $\{t_1, ..., t_n\}$. A node label s in **T** of type 3 is said to be of **type 3a** if no subterm of s has more than k arguments. Otherwise s is said to be of type **3b**.

Example 5: Let F={ a, b, f, h} with f to be an AC-function symbol. Assume t_1 = f(x,y,y) , t_2 = f(x,a), and t_3 = f(b,b). The term s= f(h(x),h(y), h(z)) is of type 3 since t is AC-unifiable with t_1 and covers it. Further, s is of type 3a since it has no more than 3 arguments. On the other hand, the term s= f(x,y,z,w) is of type 3b.

The following theorem shows that terms of type 3a are not quasi-AC-instances of a set $\{t_1, ..., t_n\}$. The correcteness proof of this theorem is based on the following arguments:

1) every variable position in the P-equivalence class of a term s of type 3a corresponds either to a linear variable in $\{t_1, ..., t_n\}$ or it is not in the domain of $\{t_1, ..., t_n\}$, and

2) There exist ground terms of any depth since the signature is extended (i.e., terms of the form f(...g(...,f(...))),...). By the first argument a ground instance of s cannot be an AC-instance of a linear-term t_j in $\{t_1, ..., t_n\}$, otherwise s would have been an AC-instance of t_j. The second argument allows to build ground instances of s of any depth and width that prevent non-linear terms in $\{t_1, ..., t_n\}$ to match them.

Theorem 4: *Let $S = \{t_1, ..., t_n\}$ be a set of terms, and let T be a complete pattern tree of a flattened term t . Leaf labels in T of type 3a are not quasi-AC-instances of $\{t_1, ..., t_n\}$.*

Let us now deal leaf labels of type 3b. The main problem with those leaf labels that there is no direct way to reason about the behavior of their ground instances. So some indirect way is needed. As we already have pointed out in the introductory example of this paper the check of whether a leaf label of type 3b is a quasi-AC-instance of S={t_1, ..., t_n} requires an *answer set* for S to be computed:

Definition 7: Let S = { t_1, ...,t_n } be a set of flattened terms. Assume that d is the depth of S and k is the maximal number of arguments of an AC-function in {t_1, ..., t_n}. The set **A(S)** = {r | r is a linear term in T(F,X) of depth bounded by d and no subterms with more than k+1 arguments and such that variables can occur only at depth d, and at the k+1th-argument of some subterm of r } is said to be an **answer set for S.**

Example 6: Let F = { f,g,a} with f to be AC. Assume that **S** is a set of terms such that depth(S) = 3 and k= 3. The set {a, g(a), g(g(x)), g(f(x,y)), g(f(x,y,z)), g(f(x,y,z,w)), f(a,a), f(a,g(x)), f(g(x),g(y)), f(a,a,a), f(a,a,g(x)), f(a,g(y),g(x)), f(g(z),g(y),g(x)), f(a,a,a,w), f(a,a,g(x),w), f(a,g(y),g(x),w), f(g(z),g(y),g(x),w) is an answer set for S. Note that the term s=f(x,y,z,w) is not in A(S) since the variables x, y, and z occur at depth 2. However, in the term s= f(g(z),g(y),g(x),w) the variable w may occur in s (depth(s)< 3) since w is the fourth argument of s.
Of course any term in the answer set for S that is an instance of another answer term can be omitted (since any ground instance of the former is also a ground instance of the latter). Further it is clear that such a set is finite and any ground term of any depth and of any width is an AC-instance of a term in A(S).

Definition 8: Let S = { t_1, ...,t_n } be a set of flattened terms, and let **A(R)** be an answer set for S. If t is a term of type 3b, then the set **I(t)$_{A(S)}$** of the instances of t which are obtained by using terms in A(S) (up to variable renaming) is said to be the **answer substitution set of t.**

When an answer set for S = { t_1, ...,t_n } is computed, the problem of determining whether a term of type 3b is a quasi-AC-instance of S can be decided as the following theorem shows :

Theorem 5: *Let S = {t_1, ..., t_n} be a set of terms, and let T be a complete pattern tree of a flattened term l. Leaf labels s in T of* *type 3b are quasi-AC-instances of { t_1, ..., t_n } iff any term in $I(s)_{A(S)}$ is an AC-instance of t_1, ...,t_n. .*

The correcteness proof of this theorem is based on the following arguments:

1) The answer-substitutions of a term s of type 3b cover the set of its ground instances. This proves the "if-part" of the theorem.

2) For the "only if- part " the existence of ground terms of any depth allows to build instances rh of r in $\mathbf{I(s)_{A(S)}}$. The terms rh have subterms that below to a constant d are all *very* different. Because of this construction rh cannot be AC-instances of any non-linear term t_j in {t_1, ..., t_n} if r is not itself an AC-instance of a term t_j . On the other hand side, because of the structure of A(S), rh is an AC-instance of a linear term t_j in {t_1, ..., t_n } iff r is an AC-instance of the term t_j .

4.2 THE (CPAC)-PROBLEM FOR BASIC SIGNATURES

Let us now consider the case of signatures with several constants and only one function of arity greater than zero **(basic signatures).** Let F be the signature {+, C_1,...,C_m} where C_i are all distinct constants, and + is an AC symbol. Let $t = \Sigma_i a_i x_i + \Sigma_k a_k C_k$, $t_1 = \Sigma_i b^1{}_i x_i + \Sigma_k b^1{}_k C_k$, ... , $t_p = \Sigma_i b^p{}_i x_i + \Sigma_k b^p{}_k C_k$, be p+1 arbritary terms, where x_i are dinstinct variables (the upper symbols 1,..,p are not power exponents but indices). Let G be the set of ground instances of t that are not instances of one of t_1,...,t_p. Elements of G are of the form $\Sigma_k c_k C_k$. We identify G with the set of vectors (c_1,...,c_m) associated to its elements. The following theorem shows that the CPAC-problem for basic signatures is decidable:

Theorem 6: *Suppose B is the maximun of the p+1 integers $\Sigma_i b^j_i + c(b^j) + sup(b^j)$ and $\Sigma_i a_i + c(a) + sup(a)$. Let D be the lcm of the p+1 integers $gcd(b^j)$ and $gcd(a)$. Then G is empty, if and only if, G u $[0; B+ D]^m$ is empty.*

The proof of this theorem is based on the following lemma:

lemma: Let $E_i = E_i' + (a_i N)^n$, i =0,....,p be p+1 subsets of N^n with $a_i \geq 0$ such that E_i is a subset of $[0;b]^n$ for all $i \leq p$. Let E = E_0 -

$\cup_{i>0} E_i$, and a = lmc(a_i). Then E is empty, iff, E u $[0; a + b]^n$ is empty.

Examples 7: a) Let F = { +, c}, and t = 2x1 + 2x2, and t1 = x1 +3x2. We then have that p = 1, a = (2,2), b1= (1,3), n=2, m =1, d(a) = 2, $d(b^1)$ = 1, D=2, c(a) = 0, $c(b^1)$ = 3, and B= 7. We can easily verify that G u [0;8] = o, so G is empty. Therefore each ground instance of t is an AC- ground instance of t1.

b) Let F = { +, c1, c2 }, t = c1 +c2 + x +y, t1 = c1+2c2+x, t2=3c1+c2, t3 = 4c1 +c2. Then B = 4 and D = 1. Therefore G is non-empty since G u $[0;5]^2 \neq$ o and the terms (5+m)c1 +c2 are the ground instances of t which are not AC-instances of t1,t2,t3,t4.

5. CONCLUSION

We have shown that the complement problem for associative and commutative theories i.e., the problem of checking whether there are ground instances of a term t which are not instances of the terms t1, ..., tn modulo sets of associativity and commutativity axioms is decidable.

This result have many implications to logic and functional programming, automated reasoning, specification analysis, and machine learning .

6. BIBLIOGRAPHY

[AS, 83] ANGLUIN,D., and SMITH, C.H: Inductive Inference: Theory and Methods *Computing Surveys, Vol. 15, No. 3, Sept. 1983.*

[COL, 84] COLMERAUER, A.: Equations and Inequations on finite and infinite trees. *Proceeding of the FGCS conference pp. 85-99 ,1984 .*

[HH,82] HUET, G. and HULLOT J.M. : Proofs by induction in equational theories with constructors. *J. Comput. System Sci. No 25-2 , 1982.*

[KNZ,86]KAPUR, D.and NARENDRAN, P. and ZHANG, H. : Complexity of Sufficient Completeness and Quasi-reducibility *Proceeding of the Conference of in Foundations of Software Technology ,1986.*

[KOU,90] KOUNALIS, E.: Pumping lemmas for tree languages generated by rewrite systems. *Proc. EUROCAL 85 , LNCS No 204, Springer-Verlag (1985) .*

[KR,91] KOUNALIS, E. and LUGIEZ, D., Compiling pattern matching with AC-functions. *Proc. 16th Conf. of CAAP, 1991 LNCS Springer-Verlag (1990)*

[LM,87] LASSEZ,J.L. and MARRIOTT, K. :Explicit representation of term defined by counter examples. *Journal of Automated reasoning 3 (1987) , pp. 301-317.*

[MAH,88] MAHER, M., Complete axiomatization of the algebra of finite, rational and infinite trees. *In Proc. of 3rd LICS, 1988.*

[MIC,83] MICHALSKI, R.S.:A theory and Methodology of Inductive Learning. *Artificial intelligence 20 (1983) , pp. 111-161*

[PJ, 87]: PEYTON-JONES: *The Implementation of Functional Programming Languages*. Prentice-Hall, 1987.

[PLO,71] PLOTKIN, G.:A Further note on Inductive Generalization . In *Machine Intelligence 6(1971)*

[SCH, 88] SCHNOEBELEN, P.: "Refined compilation of pattern-matching for functional languages". *Science of Computer Programming 11, 1988, 133-159.*

[TRE,90] TREINEN. R., A new method for undecidability proofs of first-order theories, In Proc. of FCT-TCS conference (INDIA),

Printing: Weihert-Druck GmbH, Darmstadt
Binding: Buchbinderei Schäffer, Grünstadt

Lecture Notes in Computer Science

For information about Vols. 1–610
please contact your bookseller or Springer-Verlag

Vol. 611: M. P. Papazoglou, J. Zeleznikow (Eds.), The Next Generation of Information Systems: From Data to Knowledge. VIII, 310 pages. 1992. (Subseries LNAI).

Vol. 612: M. Tokoro, O. Nierstrasz, P. Wegner (Eds.), Object-Based Concurrent Computing. Proceedings, 1991. X, 265 pages. 1992.

Vol. 613: J. P. Myers, Jr., M. J. O'Donnell (Eds.), Constructivity in Computer Science. Proceedings, 1991. X, 247 pages. 1992.

Vol. 614: R. G. Herrtwich (Ed.), Network and Operating System Support for Digital Audio and Video. Proceedings, 1991. XII, 403 pages. 1992.

Vol. 615: O. Lehrmann Madsen (Ed.), ECOOP '92. European Conference on Object Oriented Programming. Proceedings. X, 426 pages. 1992.

Vol. 616: K. Jensen (Ed.), Application and Theory of Petri Nets 1992. Proceedings, 1992. VIII, 398 pages. 1992.

Vol. 617: V. Mařík, O. Štěpánková, R. Trappl (Eds.), Advanced Topics in Artificial Intelligence. Proceedings, 1992. IX, 484 pages. 1992. (Subseries LNAI).

Vol. 618: P. M. D. Gray, R. J. Lucas (Eds.), Advanced Database Systems. Proceedings, 1992. X, 260 pages. 1992.

Vol. 619: D. Pearce, H. Wansing (Eds.), Nonclassical Logics and Information Proceedings. Proceedings, 1990. VII, 171 pages. 1992. (Subseries LNAI).

Vol. 620: A. Nerode, M. Taitslin (Eds.), Logical Foundations of Computer Science – Tver '92. Proceedings. IX, 514 pages. 1992.

Vol. 621: O. Nurmi, E. Ukkonen (Eds.), Algorithm Theory – SWAT '92. Proceedings. VIII, 434 pages. 1992.

Vol. 622: F. Schmalhofer, G. Strube, Th. Wetter (Eds.), Contemporary Knowledge Engineering and Cognition. Proceedings, 1991. XII, 258 pages. 1992. (Subseries LNAI).

Vol. 623: W. Kuich (Ed.), Automata, Languages and Programming. Proceedings, 1992. XII, 721 pages. 1992.

Vol. 624: A. Voronkov (Ed.), Logic Programming and Automated Reasoning. Proceedings, 1992. XIV, 509 pages. 1992. (Subseries LNAI).

Vol. 625: W. Vogler, Modular Construction and Partial Order Semantics of Petri Nets. IX, 252 pages. 1992.

Vol. 626: E. Börger, G. Jäger, H. Kleine Büning, M. M . Richter (Eds.), Computer Science Logic. Proceedings, 1991. VIII, 428 pages. 1992.

Vol. 628: G. Vosselman, Relational Matching. IX, 190 pages. 1992.

Vol. 629: I. M. Havel, V. Koubek (Eds.), Mathematical Foundations of Computer Science 1992. Proceedings. IX, 521 pages. 1992.

Vol. 630: W. R. Cleaveland (Ed.), CONCUR '92. Proceedings. X, 580 pages. 1992.

Vol. 631: M. Bruynooghe, M. Wirsing (Eds.), Programming Language Implementation and Logic Programming. Proceedings, 1992. XI, 492 pages. 1992.

Vol. 632: H. Kirchner, G. Levi (Eds.), Algebraic and Logic Programming. Proceedings, 1992. IX, 457 pages. 1992.

Vol. 633: D. Pearce, G. Wagner (Eds.), Logics in AI. Proceedings. VIII, 410 pages. 1992. (Subseries LNAI).

Vol. 634: L. Bougé, M. Cosnard, Y. Robert, D. Trystram (Eds.), Parallel Processing: CONPAR 92 – VAPP V. Proceedings. XVII, 853 pages. 1992.

Vol. 635: J. C. Derniame (Ed.), Software Process Technology. Proceedings, 1992. VIII, 253 pages. 1992.

Vol. 636: G. Comyn, N. E. Fuchs, M. J. Ratcliffe (Eds.), Logic Programming in Action. Proceedings, 1992. X, 324 pages. 1992. (Subseries LNAI).

Vol. 637: Y. Bekkers, J. Cohen (Eds.), Memory Management. Proceedings, 1992. XI, 525 pages. 1992.

Vol. 639: A. U. Frank, I. Campari, U. Formentini (Eds.), Theories and Methods of Spatio-Temporal Reasoning in Geographic Space. Proceedings, 1992. XI, 431 pages. 1992.

Vol. 640: C. Sledge (Ed.), Software Engineering Education. Proceedings, 1992. X, 451 pages. 1992.

Vol. 641: U. Kastens, P. Pfahler (Eds.), Compiler Construction. Proceedings, 1992. VIII, 320 pages. 1992.

Vol. 642: K. P. Jantke (Ed.), Analogical and Inductive Inference. Proceedings, 1992. VIII, 319 pages. 1992. (Subseries LNAI).

Vol. 643: A. Habel, Hyperedge Replacement: Grammars and Languages. X, 214 pages. 1992.

Vol. 644: A. Apostolico, M. Crochemore, Z. Galil, U. Manber (Eds.), Combinatorial Pattern Matching. Proceedings, 1992. X, 287 pages. 1992.

Vol. 645: G. Pernul, A M. Tjoa (Eds.), Entity-Relationship Approach – ER '92. Proceedings, 1992. XI, 439 pages, 1992.

Vol. 646: J. Biskup, R. Hull (Eds.), Database Theory – ICDT '92. Proceedings, 1992. IX, 449 pages. 1992.

Vol. 647: A. Segall, S. Zaks (Eds.), Distributed Algorithms. X, 380 pages. 1992.

Vol. 648: Y. Deswarte, G. Eizenberg, J.-J. Quisquater (Eds.), Computer Security – ESORICS 92. Proceedings. XI, 451 pages. 1992.

Vol. 649: A. Pettorossi (Ed.), Meta-Programming in Logic. Proceedings, 1992. XII, 535 pages. 1992.

Vol. 650: T. Ibaraki, Y. Inagaki, K. Iwama, T. Nishizeki, M. Yamashita (Eds.), Algorithms and Computation. Proceedings, 1992. XI, 510 pages. 1992.

Vol. 651: R. Koymans, Specifying Message Passing and Time-Critical Systems with Temporal Logic. IX, 164 pages. 1992.

Vol. 652: R. Shyamasundar (Ed.), Foundations of Software Technology and Theoretical Computer Science. Proceedings, 1992. XIII, 405 pages. 1992.

Vol. 653: A. Bensoussan, J.-P. Verjus (Eds.), Future Tendencies in Computer Science, Control and Applied Mathematics. Proceedings, 1992. XV, 371 pages. 1992.

Vol. 654: A. Nakamura, M. Nivat, A. Saoudi, P. S. P. Wang, K. Inoue (Eds.), Prallel Image Analysis. Proceedings, 1992. VIII, 312 pages. 1992.

Vol. 655: M. Bidoit, C. Choppy (Eds.), Recent Trends in Data Type Specification. X, 344 pages. 1993.

Vol. 656: M. Rusinowitch, J. L. Rémy (Eds.), Conditional Term Rewriting Systems. Proceedings, 1992. XI, 501 pages. 1993.

Vol. 657: E. W. Mayr (Ed.), Graph-Theoretic Concepts in Computer Science. Proceedings, 1992. VIII, 350 pages. 1993.

Vol. 658: R. A. Rueppel (Ed.), Advances in Cryptology – EUROCRYPT '92. Proceedings, 1992. X, 493 pages. 1993.

Vol. 659: G. Brewka, K. P. Jantke, P. H. Schmitt (Eds.), Nonmonotonic and Inductive Logic. Proceedings, 1991. VIII, 332 pages. 1993. (Subseries LNAI).

Vol. 660: E. Lamma, P. Mello (Eds.), Extensions of Logic Programming. Proceedings, 1992. VIII, 417 pages. 1993. (Subseries LNAI).

Vol. 661: S. J. Hanson, W. Remmele, R. L. Rivest (Eds.), Machine Learning: From Theory to Applications. VIII, 271 pages. 1993.

Vol. 662: M. Nitzberg, D. Mumford, T. Shiota, Filtering, Segmentation and Depth. VIII, 143 pages. 1993.

Vol. 663: G. v. Bochmann, D. K. Probst (Eds.), Computer Aided Verification. Proceedings, 1992. IX, 422 pages. 1993.

Vol. 664: M. Bezem, J. F. Groote (Eds.), Typed Lambda Calculi and Applications. Proceedings, 1993. VIII, 433 pages. 1993.

Vol. 665: P. Enjalbert, A. Finkel, K. W. Wagner (Eds.), STACS 93. Proceedings, 1993. XIV, 724 pages. 1993.

Vol. 666: J. W. de Bakker, W.-P. de Roever, G. Rozenberg (Eds.), Semantics: Foundations and Applications. Proceedings, 1992. VIII, 659 pages. 1993.

Vol. 667: P. B. Brazdil (Ed.), Machine Learning: ECML – 93. Proceedings, 1993. XII, 471 pages. 1993. (Subseries LNAI).

Vol. 668: M.-C. Gaudel, J.-P. Jouannaud (Eds.), TAPSOFT '93: Theory and Practice of Software Development. Proceedings, 1993. XII, 762 pages. 1993.

Vol. 669: R. S. Bird, C. C. Morgan, J. C. P. Woodcock (Eds.), Mathematics of Program Construction. Proceedings, 1992. VIII, 378 pages. 1993.

Vol. 670: J. C. P. Woodcock, P. G. Larsen (Eds.), FME '93: Industrial-Strength Formal Methods. Proceedings, 1993. XI, 689 pages. 1993.

Vol. 671: H. J. Ohlbach (Ed.), GWAI-92: Advances in Artificial Intelligence. Proceedings, 1992. XI, 397 pages. 1993. (Subseries LNAI).

Vol. 672: A. Barak, S. Guday, R. G. Wheeler, The MOSIX Distributed Operating System. X, 221 pages. 1993.

Vol. 673: G. Cohen, T. Mora, O. Moreno (Eds.), Applied Algebra, Algebraic Algorithms and Error-Correcting Codes. Proceedings, 1993. X, 355 pages 1993.

Vol. 674: G. Rozenberg (Ed.), Advances in Petri Nets 1993. VII, 457 pages. 1993.

Vol. 675: A. Mulkers, Live Data Structures in Logic Programs. VIII, 220 pages. 1993.

Vol. 676: Th. H. Reiss, Recognizing Planar Objects Using Invariant Image Features. X, 180 pages. 1993.

Vol. 677: H. Abdulrab, J.-P. Pécuchet (Eds.), Word Equations and Related Topics. Proceedings, 1991. VII, 214 pages. 1993.

Vol. 678: F. Meyer auf der Heide, B. Monien, A. L. Rosenberg (Eds.), Parallel Architectures and Their Efficient Use. Proceedings, 1992. XII, 227 pages. 1993.

Vol. 683: G.J. Milne, L. Pierre (Eds.), Correct Hardware Design and Verification Methods. Proceedings, 1993. VIII, 270 Pages. 1993.

Vol. 684: A. Apostolico, M. Crochemore, Z. Galil, U. Manber (Eds.), Combinatorial Pattern Matching. Proceedings, 1993. VIII, 265 pages. 1993.

Vol. 685: C. Rolland, F. Bodart, C. Cauvet (Eds.), Advanced Information Systems Engineering. Proceedings, 1993. XI, 650 pages. 1993.

Vol. 686: J. Mira, J. Cabestany, A. Prieto (Eds.), New Trends in Neural Computation. Procedings, 1993. XII, 746 pages. 1993.

Vol. 687: H. H. Barrett, A. F. Gmitro (Eds.), Information Processing in Medical Imaging. Proceedings, 1993. XVI, 567 pages. 1993.

Vol. 688: M. Gauthier (Ed.), Ada - Europe '93. Proceedings, 1993. VIII, 353 pages. 1993.

Vol. 689: J. Komorowski, Z. W. Ras (Eds.), Methodologies for Intelligent Systems. Proceedings, 1993. XI, 653 pages. 1993. (Subseries LNAI).

Vol. 690: C. Kirchner (Ed.), Rewriting Techniques and Applications. Proceedings, 1993. XI, 488 pages. 1993.

Vol. 691: M. A. Marsan (Ed.), Application and Theory of Petri Nets 1993. Proceedings, 1993. IX, 591 pages. 1993.